GENETIC ANALYSIS OF THE X CHROMOSOME
Studies of Duchenne Muscular Dystrophy and Related Disorders

ADVANCES IN EXPERIMENTAL MEDICINE AND BIOLOGY

Recent Volumes in this Series

GENETIC ANALYSIS OF THE X CHROMOSOME
Studies of Duchenne Muscular Dystrophy and Related Disorders

Edited by

Henry F. Epstein

The Jerry Lewis Neuromuscular Disease Research Center
Baylor College of Medicine
Houston, Texas

and

Stewart Wolf

Totts Gap Medical Research Laboratories
Bangor, Pennsylvania
Temple University
Philadelphia, Pennsylvania

PLENUM PRESS • NEW YORK AND LONDON

Library of Congress Cataloging in Publication Data

Main entry under title:

Genetic analysis of the X chromosome.

(Advances in experimental medicine and biology; v. 154)
"Proceedings of a colloquium sponsored by the Muscular Dystrophy Association, held December 14–16, 1981, at Mountain Shadows Inn, Scottsdale, Arizona."—p.
Includes bibliographical references and index.
1. Muscular dystrophy in children—Genetic aspects—Congresses. 2. Gene expressions—Congresses. I. Epstein, Henry F., 1944–   . II. Wolf, Stewart George, 1914–   . III. Muscular Dystrophy Association. IV. Series. [DNLM: 1. Sex Chromosomes—Ultrastructure—Congresses. 2. Muscular dystrophy—Familial and genetic—Congresses. W1 AD559 v. 154/WE 559 G328 1981]
RJ482.D9G46 1982   618.92'74   82-15037
ISBN 0-306-41129-6

Proceedings of a Colloquium sponsored by the Muscular Dystrophy Association, held December 14–16, 1981, at Mountain Shadows Inn, Scottsdale, Arizona.

©1982 Plenum Press, New York
A Division of Plenum Publishing Corporation
233 Spring Street, New York, N.Y. 10013

Printed in the United States of America

PREFACE

The present volume contains the edited transcript of a colloquium sponsored by the Muscular Dystrophy Association and held at Mountain Shadows Inn, Scottsdale, Arizona, December 14-16, 1981.

The participants, geneticists, molecular biologists, biochemists and clinicians, explored in open dialogue ways and means of identifying and characterizing the genetic alterations responsible for X-linked muscular dystrophies, especially the Duchenne type. The clinicians, who urged the use of properly diagnosed and documented case material for study, emphasized the troublesome fact that the primary phenotypic expression of the gene (or genes) involved in the muscular dystrophies is yet to be identified.

Discussions centered on the applicability of recent methodological advances in DNA chemistry and molecular biology, cytogenetics and cell biology to mapping the X chromosome. Despite ignorance of the basic disorder in the muscular dystrophies, DNA technologies and chromosome mapping strategies for the discovery of genetic defects and phenotypic expressions were proposed. Beyond its stimulating intellectual exchange, the colloquium yielded important benefits. The participants agreed to share needed cell lines and endonuclease restriction enzymes and to organize interlaboratory communication and collaborative efforts to accelerate progress in the quest for the genetic lesion in Duchenne muscular dystrophy.

The discussions were recorded, transcribed, edited and to some extent, rearranged to fit into a sequence of chapters. The editors are grateful to Joy Colarusso Lowe whose unusual skill, patience and persistence made it possible to convert a highly specialized technical discussion into a coherent manuscript.

PARTICIPANTS

Samuel H. Boyer, IV, M.D., Department of Medicine, Johns Hopkins
    University School of Medicine, Baltimore, Maryland.

Michael H. Brooke, M.D., Department of Neurology, Washington
    University School of Medicine, St. Louis, Missouri.

Gail A.P. Bruns, M.D., Department of Pediatrics, Children's
    Hospital Medical Center, Harvard Medical School, Boston,
    Massachusetts.

C. Thomas Caskey, M.D., Departments of Medicine and Biochemistry,
    Baylor College of Medicine, Howard Hughes Medical Institute,
    Houston, Texas.

A. Craig Chinault, Ph.D., Departments of Medicine and Biochemistry,
    Howard Hughes Medical Institute, Baylor College of Medicine,
    Houston, Texas.

P. Michael Conneally, Ph.D., Department of Medical Genetics,
    Indiana University School of Medicine, Indianapolis, Indiana.

Norman Davidson, Ph.D., Department of Chemistry, California Institute
    of Technology, Pasadena, California.

Richard Doherty, Ph.D., Departments of Pediatrics, Obstetrics and
    Gynecology, University of Rochester School of Medicine,
    Rochester, New York.

Charles P. Emerson, Jr., Ph.D., Department of Biology, University
    of Virginia, Charlottesville, Virginia.

Henry F. Epstein, M.D., Department of Neurology, Baylor College of
    Medicine, Houston, Texas.

Uta Francke, M.D., Department of Human Genetics, Yale University
    School of Medicine, New Haven, Connecticut.

James F. Gusella, Ph.D., Genetics Unit, Massachusetts General Hospital, Harvard Medical School, Boston, Massachusetts.

Stephen D. Hauschka, Ph.D., Department of Biochemistry, University of Washington, Seattle, Washington.

Robert E. Hausman, Ph.D., Biological Science Center, Boston University, Boston, Massachusetts.

Barbara K. Hecht, Ph.D., Genetic Center, Southwest Biomedical Research Institute, Tempe, Arizona.

Frederick Hecht, M.D., Genetic Center, Southwest Biomedical Research Institute, Tempe, Arizona.

David Housman, Ph.D., Center for Cancer Research, Massachusetts Institute of Technology, Cambridge, Massachusetts.

Oliver W. Jones, M.D., Department of Medicine, University of California, San Diego, La Jolla, California.

Simona Kahane, Ph.D., Sloan-Kettering Institute for Cancer Research, New York, New York.

Laurence H. Kedes, M.D., Veterans Administration Hospital, Palo Alto, California.

Louis M. Kunkel, Ph.D., Division of Genetics, Children's Hospital Medical Center, Harvard Medical School, Boston, Massachusetts.

Samuel A. Latt, M.D., Ph.D., Department of Pediatrics, Children's Hospital Medical Center, Harvard Medical School, Boston, Massachusetts.

Anthony N. Martonosi, M.D., Department of Biochemistry, State University of New York, Syracuse, New York.

M.L. Moss, Ph.D., Director of Research Development, Muscular Dystrophy Association, New York, New York.

Bernardo Nadal-Ginard, M.D., Ph.D., Department of Cell Biology, Albert Einstein College of Medicine, Bronx, New York.

Robert L. Nussbaum, M.D., Department of Medicine, Howard Hughes Medical Institute, Baylor College of Medicine, Houston Texas.

Bruce M. Paterson, M.D., National Cancer Institute, National Institutes of Health, Bethesda, Maryland.

Everett C. Schreiber, Jr., Ph.D., Research Department, Muscular
    Dystrophy Association, New York, New York.

Robert J. Schwartz, Ph.D., Department of Cell Biology, Baylor
    College of Medicine, Texas Medical Center, Houston, Texas.

Lawrence J. Shapiro, M.D., Division of Medical Genetics, Harbor
    General Hospital, Torrance, California.

Marcello Siniscalco, M.D., Somatic Cell Genetics, Sloan-Kettering
    Institute for Cancer Research, New York, New York.

Kirby D. Smith, Ph.D., Department of Medicine, Johns Hopkins
    University School of Medicine, Baltimore, Maryland.

Richard C. Strohman, Ph.D., Department of Zoology, University of
    California, Berkeley, Berkeley, California.

Paul Szabo, Ph.D., Somatic Cell Genetics, Sloan-Kettering Institute
    for Cancer Research, New York, New York.

Ray White, M.D., University of Utah School of Medicine, Howard
    Hughes Medical Institute, Salt Lake City, Utah.

Robert Williamson, M.D., Ph.D., Department of Biochemistry, St. Mary's
    Hospital Medical School, University of London, United Kingdom.

Stewart Wolf, M.D., Director, Totts Gap Medical Research Laboratories,
    Bangor, Pennsylvania; Professor of Medicine, Temple University
    School of Medicine, Philadelphia, Pennsylvania.

CONTENTS

GLOSSARY

In the discussions, a number of recently introduced or highly
technical genetical terms were used. Definitions from a general
biomedical viewpoint are provided.

BAYES' THEOREM

A theoretical procedure by which the ordinary probability is
altered by knowledge of relevant outcomes leading to a conditional
probability. For example, a woman who has a one-half chance of
being a carrier for an X-linked disorder, à priori, has with each
succeeding normal male child in the absence of any affected male
child a decreasing conditional probability of actually being a
carrier.

cDNA

cDNA is a complementary DNA copy of a specific messenger RNA
(mRNA) or of that nucleotide sequence actually coding for a specific
sequence of amino acids that forms a biological polypeptide. Originally, cDNA was produced by mRNA by the avian virus enzyme, reverse
transcriptase. Such cDNAs can be cloned by recombinant DNA methods
and can be used as probes for quantitating either mRNA content or
chromosomal genes. The practical equivalent of a cDNA can now be
obtained from appropriate clones of genomic DNA.

CHO

Chinese hamster ovary cells are a very well studied line that
has been particularly suitable for somatic cell genetic studies.

CK - creatine kinase

The enzyme which produces the phosphagen, creatine phosphate,
that in turn buffers the levels of ATP in cells, especially striated
muscles. In vertebrates and humans, the general form of the enzyme
is a BB dimer. In cardiac and developing skeletal muscles, the
MB dimer is formed whereas in differentiated skeletal muscle, the
MM dimer is found exclusively. Increased levels of CK activity are
found in Duchenne muscular dystrophy patients, and in some of their
family members as well as in other pathological and physiological
states.

cM - centimorgan

Equivalent to one percent recombination of two genetic markers
due to crossover between two homologous chromosomes during meiosis.

The term honors Thomas Hunt Morgan, the father of experimental genetics, who discovered genetic linkage and related it to chromosome
location.

<u>COLCHICINE</u>

A drug that inhibits mitosis by disrupting the microtubules
of the mitotic spindle. The compound binds specifically to the
major protein subunit of microtubules, tubulin. Many other processes
that require microtubules including the growth and function of nerves
can be inhibited by colchicine. The drug is useful in eliminating
symptoms of gout and Mediterranean fever.

<u>CYTOCHALASIN B</u>

One of a series of compounds that inhibit cytokinesis or the
cleavage of cells in cell division. Cytochalasin B affects the
polymerization of actin in many cells and <u>in vitro</u>.

<u>FACS</u> - fluorescence-activated cell sorter

A type of instrument using laser beams that detects the fluorescence or absence of fluorescence of individual microscopic particles
and sorts them according to this detection. The particles may be
previously separated by flow through a specific tube. Cells may be
separated by their binding.

<u>GENETIC MARKERS</u>

A set of enzymes showing polymorphisms or clinical syndromes
with inherited bases that are useful as markers of specific genetic
loci on particular chromosomes.

ACP	-	acid phosphatase (autosomal, 11)
AD	-	adenosine deaminase (autosomal, 20)
ALD	-	adrenoleucodystrophy (X-linked)
APRT	-	adenine phosphoribosyltransferase (autosome)
CBD	-	colorblindness, deutan (X-linked)
CBP	-	colorblindness, protan (X-linked)
CGD	-	chronic granulomatous disease (X-linked)
GALA	-	α-galactosidase A (X-linked)
HEMA	-	hemophilia A (X-linked)
LDH	-	lactic acid dehydrogenase (autosomal, 11)
OA	-	optic atrophy (X-linked)
PFK	-	phosphofructokinase (autosomal, several chromosomes)
PGK	-	phosphoglycerokinase (X-linked)
PRPS	-	phosphoribosyl pyrophosphate synthetase (X-linked)
RS	-	retinoschisis (X-linked)
STS	-	steroid sulfatase (ichthyosis) (X-linked)
Xg(a)-		blood group (X-linked)

<u>G6PD</u> - glucose 6-phosphate dehydrogenase

This is a key enzyme in the pentose phosphate shunt of glycolysis that produces NADPH. There are normal variants (A^+ and B) as well as clinically significant mutant enzymes. The locus of the enzyme is X-linked.

<u>HeLa</u>

A well studied human cell line originally derived from a carcinoma of a patient named Helen Lane. These cells are a common source of human DNA although their aneuploid nature suggests the possibility that the DNA may not be representative.

<u>HPRT</u> - hypoxanthine-guanine phosphoribosyltransferase

This enzyme is necessary for the salvage of these purine bases to produce the appropriate nucleoside monophosphates. There is a severe form of the enzyme deficiency in male children called <u>Lesch-Nyhan</u> syndrome, and a milder form in adults associated with gout. The enzyme locus is X-linked. Cells deficinet in HPRT are resistant to killing doses of <u>8-azaguanine</u>.

<u>kb</u> - kilobases

One thousand base pairs or a multiple thereof of double-stranded DNA. May refer to bases in single-stranded DNA or RNA as well.

λ - bacteriophage lambda

A bacteriophage that has been widely used in genetics and molecular biology that becomes incorporated into the <u>E. coli</u> chromosome. Certain modified λ phages such as <u>gtwes</u> and <u>Charon 4A and 21A</u> are widely used in recombinant DNA work because they contain sites for specific endonucleases. <u>Cosmids</u> are modified λ phages that can accept pieces of DNA up to 40 kb.

<u>LIBRARY</u>

A collection of either genomic or cDNA sequences cloned in a particular vector that represents the genetic information of a species or the transcribed information of a cell population, respectively.

LOD

Logarithm (base 10) of the ratio of likelihood of one model
to the likelihood of a second model (likelihood ratio). A LOD
score of 0 is equal likelihood, of 3 is a 1000-fold ratio or
"certainty" of one model over another. The LOD score is used in
genetics as an index of linkage between two markers, that is the
likelihood of finding the two linked over unlinked in a particular
group of individuals.

LYONIZATION

An hypothesis suggested by M. Lyon to explain the observed
pattern of X-linked gene expression of females and in males with
multiple X chromosomes. Random inactivation of either X chromosome
occurs in all cell lineages, leaving the other X active in terms
of specific gene expression. Certain abnormal X chromosomes such
as X-autosome translocations may remain active preferentially, and
in those cases, the normal X is inactive.

μ, ν

The forward rates of mutation of gametes in males and females.
Each character or gene may be assigned such a rate, and any set of
genes may also have an average rate. Sites within genes may have
different rates too. It is important to note that the rates of
mutation in sperm or egg production may be different, and assumptions
that they are the same may be significantly incorrect.

NICK TRANSLATION

A set of methods for radioactively labeling DNA molecules
in vitro by the action of a DNA polymerase.

PLASMID

A circular double-stranded DNA molecule capable of autonomous
replication in the cytoplasm of specific bacteria. Plasmids have
been very useful in forming recombinant DNA molecules with DNAs
from a variety of sources. pBR322 is a specific plasmid that can
infect Escherichia coli which is widely used because of internal
sites that can be cleaved by specific endonucleases, and drug
resistant genes that permit ready selection.

REPETITIVE DEFICIENT

A series of cloned DNA sequences from human sources generated
by Alu I action. Fragments so produced generally contain repetitive
sequences about the Alu site. These fragments have been tested for
absence of such repetitive sequences and may have been treated with

other restriction endonucleases to remove the repetitive sites.
<u>Blur 11</u> is a specific cloned <u>Alu I</u>-generated human DNA sequence.

<u>RESTRICTION ENDONUCLEASES</u>

Enzymes that make internal breaks at highly specific base
sequences in double-stranded DNA. These sites of restriction are
usually pallindromic in that the sequence on one strand is the
reverse of the sequence on the second strand. Restriction enzymes
are produced by specific strains of bacteria as protection against
infection by viruses. Particular enzymes mentioned during discussions, their source and their specificity in terms of base sequence
are as follows:

EcoRI	<u>Haemophilis influenzae Rd</u>	A ↓ AGCTT
Alu I	<u>Arthrobacter luteus</u>	AG ↓ CT
Bam HI	<u>Bacillus amyloliquefaciens H</u>	G ↓ GATCC
SAC I	<u>Streptomyces achromogenes</u>	GAGCT ↓ C
Msp I	<u>Moraxella species</u>	CC ↓ GG
Hpa I	<u>Hemophilus parinfluenzae</u>	GTT ↓ AAC
Taq I	<u>Thermus aquaticus</u>	T ↓ CGA
Bgl I	<u>Bacillus globigii</u>	GCC(N₄) ↓ NGGC
		(N = any base)
Bgl II	<u>Bacillus globigii</u>	A ↓ GATCT
Kpn I	<u>Klebsiella pneumoniae</u>	GGTAC ↓ C
*Mbo I	<u>Moraxella bovis</u>	↓ GATC
*Sau 3A	<u>Staphylococcus aureus 3A</u>	↓ GATC
Xba I	<u>Xanthamonas badrii</u>	T ↓ CTAGA

*Isoschizomers – Enzymes that can cut at
the same sequence.

<u>RFLP</u> – restriction fragment length polymorphism

Inherited polymorphism or variant that alters the pattern
of restriction endonuclease digestion of DNA. The mutation could
affect the restriction site directly or cause an insertion or
deletion in DNA near the site.

<u>SOUTHERN</u>

Refers to a procedure introduced by E. Southern of transferring
pieces of DNA (usually generated by the action of specific endonucleases) from a gel electrophoretic separation to nitrocellulose
paper. The resulting blot can be treated with radioactive DNA
probes to detect the present of specific sequences within the
generated fragments. In a <u>Northern,</u> RNA molecules are separated
and then transferred. In a <u>Western</u> protein molecules are separated,
transferred and identified by their reaction with specific antibodies.

The <u>Benton-Davis</u> and <u>Grunstein-Hogness</u> procedures use nitrocellulose paper to immobilize DNA from bacteriophage-induced plaques or plasmid-infected bacterial colonies, respectively, which are then treated with a specific readioactive DNA probe.

<u>SV40</u> - Simian virus 40

A virus that can transform a variety of cell types and produce permanent cell lines that in animals form tumors. The chromosome of SV40 is very small, and the structure and function of its DNA has been determined in detail. Modified hybrids of SV40 and λ phage can be sued as universal vectors for DNA transfer, either into animal or bacterial host cells. The SV40 promoter or initiation site for mRNA synthesis can function with DNA inserts from other sources and lead to the synthesis of exogenous proteins in appropriate animal or bacterial hosts.

<u>TK</u> - thymidine kinase

The enzyme catalyzes the production of dTMP (deoxythymidine monophosphate) from thymidine and represents one pathway of dTTP (deoxythymidine triphosphate) biosynthesis, necessary for DNA synthesis and cell division. Cells deficient in this enzyme are resistant to killing doses of BrdU (5 bromodeoxyuridine), a thymidine analogue. TK^- cells are frequently used in genetic and transfection experiments with <u>BrdU</u> as the selecting drug.

<u>VECTOR</u>

An agent consisting of a DNA molecule that can replicate automomously the host cell's chromosomes and to which another DNA segment may be attached and also be replicated.

<u>WALKING</u>

The spanning of large distances of chromosomal DNA by the molecular overlapping of cloned DNA sequences.

<u>Xp21</u>

X chromosomes, short arm, band 2, subband 1. The same formalism could be used for other chromosomes and their regions. For autosomes a number instead of X would be used. <u>p</u> or <u>q</u> refer to short or long arm. The band and subband designations are usually the result of <u>high resolution prophase chromosome band</u> patterns following <u>trypsin</u> treatment and <u>Giemsa</u> staining.

CHAPTER 1

PREVALENCE AND HERITABILITY OF DUCHENNE
MUSCULAR DYSTROPHY

DR. P. MICHAEL CONNEALLY: Duchenne muscular dystrophy is inherited as an X-linked recessive disorder. The lesion seems to be
in the middle of the short arm of the X chromosome. There is also
an autosomal recessive condition which is very akin to Duchenne and
is relatively common among the Amish (1). For practical purposes,
Duchenne dystrophy occurs only in males. Although cases of
females with Duchenne muscular dystrophy have been reported (2),
the majority are Turner's syndrome (XO) or structural abnormalities
of the X. In fact, you will see stated in textbooks that the
frequency of Duchenne muscular dystrophy in Turner's is the same
as the frequency in males because they only have one X chromosome.
This is not quite true. If, for example, all of the nondisjunction
occurs in the male, then the stated frequency in Turner's syndrome
would be correct. If, on the other hand, all of the nondisjunction
occurred in the female, then all of the X's would come from the
male; in this case you would have no Duchenne since it is an X-
linked lethal. There is also a possibility of extreme lyonization
which could also cause Duchenne in females if, by chance, all or the
vast majority of their normal X's are inactivated (2).

Figure 1 shows a map of the X. The Duchenne gene is generally
thought to be on the short arm of the X and it is not closely linked
to any of the known X-linked markers; for example, the Xg blood group.
Duchenne is one of the most common X-linked disorders with a frequency
of at least 1 in 4800 males.

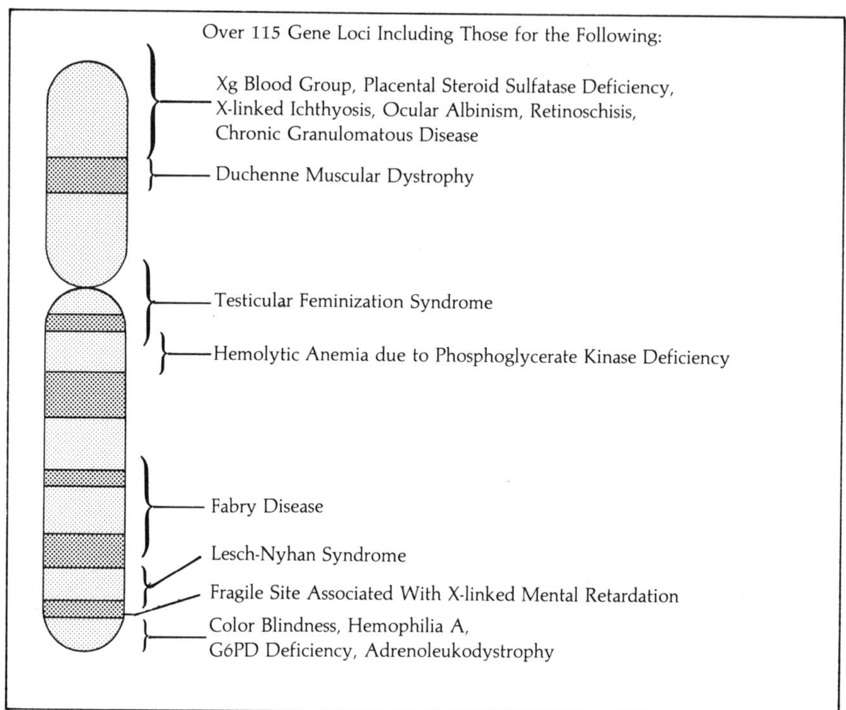

Figure 1: Linkage map of the X chromosome. From: McKusick, V.A., The Anatomy of the Human Genome; Hospital Practice, p. 85, 1981.

There are a number of ways to determine the frequency of carrier females in the population. One way to become a carrier for the Duchenne gene is through a mutation in the parental sperm at a rate per gamete in males (μ)

Carrier Females

or in the egg at rate per gamete in females (ν). The overall mutation rate to produce a carrier female is $\mu + \nu$. The mutation could have occurred in her mother's father or in her mother's mother and then be transmitted to her at a rate of 1/2 ($\mu + \nu$). The mutation could also have occurred one generation previous to this with a probability of one-fourth that it is transmitted two generations, etc. The overall probability that the woman is a carrier is ($\mu + \nu$) + 1/2 ($\mu + \nu$) + 1/4 ($\mu + \nu$) + 1/8 ($\mu + \nu$), etc. This is the sum of an infinite number of terms of a geometric series and becomes 2 ($\mu + \nu$) (3).

If the mutation rate is the same in males and females, ($\mu=\nu$), then the frequency of carrier females in the population

is 4µ. In carrier females one-half of the Duchenne genes are
new mutations and a further one-fourth are only one generation
"old." For example, $(1/2)^{10}$ or only one-tenth of one percent have
existed ten or more generations.

There are two ways to get an affected male: one from a carrier
female and the other from a new mutation. The chance that the mother
is a carrier is 4µ and the chance of transmitting the gene to her
son is 2µ. Adding the probability of a new mutation, µ, the frequency of carrier males in the population becomes 3µ. This is called
Haldane's rule, after J.B.S. Haldane, the great English population
geneticist (4). Assuming a stable frequency of affected males
and recognizing that they will not reproduce, one-third of all cases
of Duchenne would be new mutations, and two-thirds would be due to
preexisting genes. Therefore, a substantial number of mothers of
affected individuals should themselves not be carriers, and their
recurrence risk is essentially zero, i.e. the risk should only be
the mutation rate. The frequency of cases of DMD is 1 in 4800, and
theoretically this is three times the mutation rate. Therefore,
dividing by 3, we find a mutation rate of 1 in 15,000 or approximately
7 in 100,000. That is a very high mutation rate, higher than most
other estimated mutation rates in man.

What is the most plausible explanation for the high frequency
of the disease if one does not accept this high mutation rate? One
possible reason is genetic hetero-
Genetic Heterogeneity and   geneity, that is more than one locus
Mental Retardation   is involved in Duchenne. The prime
example of such a hypothesis is
xeroderma pigmentosum where there are 7 or 8 complementation groups
with distinct genetic lesions known in most of them. Clinical heterogeneity may be found in Duchenne. If a patient is severely mentally
retarded and has an affected sib, his sib will also be retarded.
Normal and retarded Duchennes are not found in the same family (5).
This suggests that there is genetic heterogeneity, expressed as
Duchenne with and Duchenne without severe mental retardation.

DR. BROOKE: We have been studying 150 boys with Duchenne
dystrophy followed in four separate university clinics. Apart from
the Becker type which we separated out, we found interesting evidence of heterogeneity unrelated to IQ. We called these cases
outliers because they are doing better than the others and not
deteriorating as rapidly. We identify them by the strength of the
neck flexors and by the creatine in the urine, but not by the IQ.
We found mixtures in the same family of bright and dull, instances
where, for example, one Duchenne child is living with the father
ar⁴ the other is living with the mother and in that situation we
see differences in IQs. We see no correlation between IQ and
muscle strength in our series.

DR. WILLIAMSON: Is there any similarity clinically or any difference in the form of mental retardation found associated with Duchenne as compared to X-linked mental retardation?

DR. CONNEALLY: I will ask Dr. Hecht to comment on that.

DR. HECHT: The fragile site on the X chromosome is between bands q27 and q28.   The segregation is as you would expect in an X-linked trait, with one-half males expressing mental retardation but only a small percentage of females.   When the females are mentally retarded, they have milder manifestations.   There is an argument over whether there is a consistent somatic phenotype associated with the retardation.   A high proportion of the affected males definitely have macroorchidism (large testes).   There may also be other consistent clinical features, large ears, for example.   There are at least two other forms of X-linked mental retardation which have not been as well studied so far as that associated with the fragile X.

DR. SINISCALCO: My colleagues, Dr. G. Filippi of the University of Trieste and Dr. A. Renaldi, University of Cagliari studied families segregating for two types of X-linked retardation and G6PD deficiency, the one with macroorchidism and the fragile site at the Xq27-Xq28, and the so-called Renpenning syndrome without macroorchidism or fragile site.   Neither of these types showed linkage with the G6PD locus.   Thus, the location of the fragile site at the tip of the X chromosome long arm does not necessarily imply that a locus for X-linked mental retardation is located in the same chromosomal region.

DR. HECHT: What is the location of the X fragile site as regards glucose 6-phosphate dehydrogenase (G6PD) and hypoxanthine-guanine phosphoribosyltransferase (HPRT).   Are they in that order?

DR. FRANCKE: The precision of mapping of HPRT and G6PD is really not to that level.   G6PD is more towards the distal end. It is in the terminal band and HPRT is in the subterminal band. How they relate to the fragile site is not precisely known.

DR. WILLIAMSON: Does the mental retardation, when it accompanies DMD, resemble that associated with X-linked fragile site?

DR. FRANCKE: Is there a difference in the degree of retardation, the IQ range, between DMD and the fragile X syndrome?

DR. BROOKE: The average IQ of our kids in this study is 105. Since we are involved in a long 4-5 year multiclinic trial, we

tend to select the kids that are going to cooperate and hence are
more intelligent. In that sense we have a bad selection. Most of
the good studies simply show that the curve is shifted to the left,
with an average IQ somewhere between 80-85.

I know nothing about other forms of mental retardation. The
kids with Duchenne are very often hyperactive and have spatial
problems. I wonder if that is true of the fragile X children.

DR. HECHT: The children with the fragile X that I have seen
are likewise hyperactive. My colleague, Dr. Barbara Hecht, just
made a point to me that it would be worthwhile in such families
with Duchenne and mental retardation segregating together to do
studies with the fragile X.

DR. WILLIAMSON: If what Dr. Brooke said is right, that the IQ
curve in DMD shows a normal distribution centered around 85, surely
it differs from fragile site X linked mental retardation where the
IQ is usually below 75.

DR. HECHT: I agree completely. I think that the mental retardation associated with the fragile X is usually around IQ 40-50.

DR. WOLF: Is there a possibility that the people studied for
the fragile X were the people most clearly mentally deficient?

DR. HECHT: Even if you throw them out, even if you throw out
the index cases, the affected sibs and maternal uncles and so forth
are really quite consistent as regards IQ. It is moderately
severe mental retardation, and it is quite consistent.

DR. WOLF: Dr. Conneally, you said that in families with more
than one child with Duchenne either all will be severely retarded
or none will be. What is the evidence? Also does mental retardation occur as a feature of any of the Becker cases?

DR. CONNEALLY: Emery, Skinner and Holloway state, "In our
experience whenever an index case is severely mentally handicapped
and has an affected brother, the latter is also severely mentally
handicapped" (5). They reviewed five studies including that of
Prosser et al, whose study mirrors what Dr. Brooke just said, that
the curve is shifted to the left (6). The IQ of their patients
overall was 87 and 30 percent of them had IQs less than 75. They
did not mention this familial propensity in their case and to my
knowledge the only one that does is Emery et al who also mentions
that about 1/3 of IQs in the 15 studies were less than 75. They
do not mention Becker; I suspect if they found mental retardation
in Becker they would have mentioned it. Again, the Prosser et al

paper says, "In those cases of Duchenne that they found were of low
mentality, that typically, the parents were too" (6). They did
an analysis of variance and found it was highly significant.

There are a number of other reasons to suspect genetic heterogeneity. Samaha and his co-workers found two different patterns of
proteins in muscle sarcoplasmic reticulum in Duchenne, and Roses
and his group at Duke found different patterns in spectrin polypeptides (7, 8). It is obviously very critical for those
who are now working on linkage and on the restriction fragment
polymorphisms to recognize that there may well be heterogeneity.

Let us return to the situation where 1/3 of all Duchennes are
new mutations, and, therefore, that their mothers are normal.

New Mutations — Gruemer's work on capping (9) and Roses' on phosphorylation of the erythrocyte membrane proteins (8)
suggested that a high percentage of mothers than expected were
carriers. A plausible explanation of the large number of Duchenne
carriers among mothers could lie in a mutation rate that differed
between males and females. It would have to be about eight times
higher in males than in females. Presumably the mutation occurred in
the maternal grandfather's sperm, making the mother the carrier of
a new mutation. There is evidence from some autosomal dominant
disorders that mutations in sperm occur at a high frequency among
the old fathers.

Some surveys have yielded data consistent with Haldane's rule.
Both Yasuda and Kondo in Japan (10) and Sibert et al in Wales
(11) found that approximately a third of all sporadic cases of
Duchenne dystrophy had normal mothers by their criteria.

DR. HOUSMAN: Are the statistics for various populations
similar or is there any variation in the frequency of Duchenne
among different populations?

DR. CONNEALLY: There are no significant population differences
in frequency. There is, however, a lot of underascertainment;
Emery et al point out that the frequency of 1 in 4800 is probably
low (5).

DR. NUSSBAUM: Haldane's rule should apply to all cases of
muscular dystrophy not just sporadic cases (4). The sibship size
has an effect on the percent of sporadic cases that are due to new
mutations. The percent of sporadic cases due to new mutations
actually increases as the sibship size increases.

DR. CONNEALLY: Yes, this is correct. Since affected males do not reproduce, the number of potential new cases of Duchenne dystrophy would be reduced by one-third, therefore, to maintain the existing prevalence of the disease would require a similar number of fresh mutations.

DR. LATT: Independent of the ratio of μ to ν in an X-linked lethal, fully half of all heterozygous females will represent fresh mutations. So obviously there is an internal check on those data, but has such a study been done with capping? If it is a questionable assay, we shouldn't build too much on it.

DR. CONNEALLY: I'm not sure. Gruemer is very dogmatic (9). He simply says the reason others can't do it is that they don't have the proper techniques. He did blind studies and I understand that he has correctly identified carriers in blind studies. Can anyone verify that?

DR. MOSS: Drachman reported that blinded samples sent from Baltimore to Gruemer were correctly identified. In addition, there is the work of Shapiro which is in press which seems to confirm the original findings. I think the upshot of it is that there is something to it. I don't believe anyone says that it is a test that can be routinely used by any laboratory that hasn't been trained by the people that originated it. I think it is still unsettled.

DR. EPSTEIN: If you accept the assertion that there is no single method or combination of methods that can reliably detect carriers, then you will miss some percentage of carriers. Is there any other information that would suggest a high mutation frequency or genetic heterogeneity independent of any one carrier detection test?

DR. CONNEALLY: The occurrence of mental retardation in some cases seems to offer evidence of heterogeneity.

DR. BOYER: Associated mental retardation may not be evidence for genetic heterogeneity, but rather for genetic modifiers. For example, all cases of sickle-cell
Genetic Modifiers   anemia have the same mutation but may
                                  vary greatly in clinical manifestations.
The same fundamental β^S mutation may have arisen at different times. Although they look very, very different, the differences are explained by differences in the rest of the genome. The clinical evidence that you cite for genetic heterogeneity may be nothing more than differences in the genetic background in which the primary locus is expressed. The fundamental mutation rate in

humans is estimated by different people at around 10^{-8} mutations
per nucleotide per year. That is not so great. The mutation rate
at the β-globin locus is around 10^{-5} per generation. Although
Duchenne may be high, it isn't as high as we thought high was. It
may be that it is a very protected locus and that any change that
you exert on it gives trouble.

There are ways that Haldane's rule could be circumvented.
Duchenne could be a heterogeneous condition because some of the
mutations in some populations may be selected in the heterozygous
state. Selection is difficult to prove but it may be that some of
these mutations have selective advantages. New mutations that are
not conveying any selective advantage because one big block of the
chromosome is gone may be lethal in utero so we never score that
as Duchenne.

DR. SINISCALCO: It seems to me an analysis of the modifiers
that govern the expression of traits within pedigrees is important.
As Haldane taught us, if there are multiple alleles you should not
have much variability within the pedigree. We recently have examined
the question of multiple alleles and modifiers in hemophilia. I
don't know whether the creatine kinase (CK) situation may be similar
to the Factor VIII deficiency. In hemophilia there is a mutation
which is X-linked, but the actual genes which are responsible for
Factor VIII synthesis are autosomal. We found a significantly lower
variance within sibships than between sibships. I don't know
whether this would be true of CK, but I wonder if anyone has ever
tested the normal brothers throughout DMD pedigrees.

DR. CASKEY: Dr. Nussbaum and I studied the families of some
42 cases of sporadically occurring Duchenne muscular dystrophy.
We used essentially the same diagnostic criteria for Duchenne that
Dr. Brooke and his colleagues use.

Problems of Carrier
Detection

We considered them pretty accurate,
at least 90 percent or better, and yet in 23 of those 42 families
we had no evidence for carrier females.

DR. NUSSBAUM: Since the CK test yields significant numbers
of false positives, instead of labeling mothers as carriers or
non-carriers on the basis of a single CK value elevated in another
female in the pedigree, we calculated the probability by Bayes'
theorem (3). In sufficiently large pedigrees with a single case
of DMD, even when the mother may be correctly scored as a non-
carrier on the basis of CK, a high CK in another female in the
pedigree is probably a false positive.

DR. CASKEY: I would urge the molecular biologists to accept that there is a high frequency of Duchenne occurring <u>de novo</u> and to consider possible molecular models that may lead to an experimental design for the identification of the actual genes involved in the Duchenne muscular dystrophy locus. There may be some clinical leads here which the molecular biologist may be able to make use of in designing specific experiments.

DR. EPSTEIN: You are saying there were multiple families with multi generational pedigrees in which there were no detectable carrier females?

DR. CASKEY: Yes.

DR. CONNEALLY: The difficulty of carrier detection is not confined to Duchenne. For example, there are multiple ways to detect carriers for cystic fibrosis, but each of them can only be done in one or two laboratories.

Figure 2 shows comparative concentrations of CK, still the most widely accepted method of carrier detection in Duchenne. You will notice that the controls have a much lower CK level than do obligate carriers, but there is a wide variance. Until we can work with a clone of cells and circumvent the lyonization problem, we will continue to have a problem with any test of carrier detection in Duchenne (12).

DR. BROOKE: The measurement of creatine kinase is helpful in diagnosis, but it is far from being a genetic marker. Creatine kinase may be increased by inflammatory diseases of muscle and by vigorous exercise in normal people, especially in those unaccustomed to exercise. Three minutes of weightlifting elevate the creatine kinase to about as much as an hour and a half of rather gentle bicycling.

The Origin and Significance of Creatine Kinase

DR. WOLF: Dr. Brooke, do you have clear-cut evidence that individuals in poor physical condition who exercise have an outpouring of CK whereas people who are in training who expend the same amount of energy do not?

DR. BROOKE: Yes. In our training studies we showed that the exercise-provoked elevations of CK were obliterated by a four week period of physical training.

DR. CONNEALLY: There are a number of things that need to be thought of in carrier detection, such as an effect of age on CK (Figure 3) as well as the effect of exercise, as mentioned by Dr. Brooke.

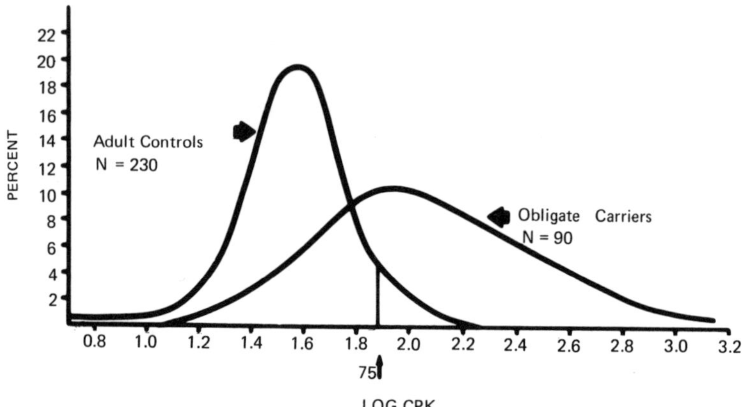

Figure 2: Distribution of CK levels in obligate
carriers of DMD and adult controls.

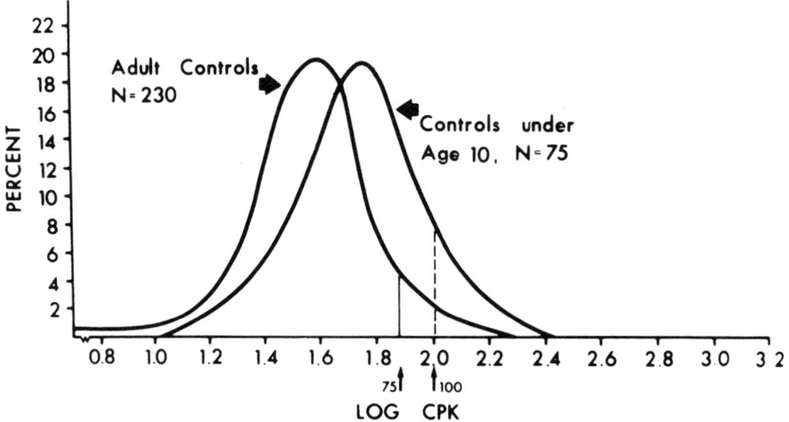

Figure 3: Distribution of CK levels in adult
controls versus controls under age 10.

Finally, let us consider the carrier probability in counseling.
Figure 4 illustrates the case of an individual, Cathy (3). She
is the daughter of a known carrier and she has half a chance at
birth of being a carrier. Later, she has a normal son and this
affects her probability; it is now no longer one-half. This type
of pedigree is the basis for the use of Bayes' theorem in genetic
counseling in X-linked disorders. If we take 200 females like Cathy,
at birth 100 of them were normal and 100 were carriers. Now assume
that they each have a son. The 100 normal mothers will all produce
normal sons, 50 of the 100 carriers will produce a normal son and
50 will produce an affected son. Our total unknown population has
been reduced by virtue of the fact that we now know that the 50
with affected sons are carriers so our total population is no longer
200, but 150 among whom 50 of them are carriers. So our probability
that Cathy is a carrier is no longer one-half but has been reduced
to one-third. The probability can be obtained using Bayes' theorem
as shown in Figure 5 (3).

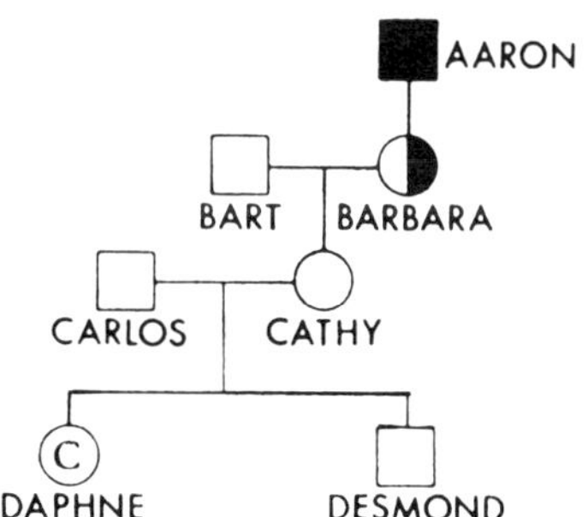

Figure 4: Pedigree of "Cathy" for Bayes'
theorem example.

State of Nature	Cathy a carrier	Cathy not a carrier
Prior probability	1/2	1/2
Conditional probability	1/2	1
Joint probability	1/4	1/2
Posterior probability	1/3	2/3

Figure 5: Calculations for example from Figure 4.

The posterior probability is simply normalizing, getting the joint
probabilities to add to 1. We find her probability is one-third
of being a carrier, two-thirds of not being one. If she had two
normal sons it would be one-fifth and four-fifths, three normal
sons would be one-ninth and eight-ninths and so on.

DR. CASKEY: Does anyone have any lead as to why the CK is
elevated in Duchenne dystrophy-affected males or carrier females?
However unreliable, it is our best biochemical index at this point.

DR. BROOKE: I am not going to be able to explain why the CK
goes up, but I can tell you some interesting aspects. One of our
youngest patients with Duchenne dystrophy was three days of age,
and we have several under the age of one. It has long been thought
that the CK is highest at the earliest part of the disease and
thereafter progressively declines. In the very youngest patients
there is some evidence that the CK is progressively higher during
the first months of life. Perhaps the CK is elevated because these
children are beginning to exercise muscles that can't take it and
the elevation of CK is equivalent to that seen in an untrained state.

Myoglobin is also released after exercise. While after exercise
the rise in CK peaks between 12 and 18 hours, myoglobin peaks
between two to three hours. I am suggesting that perhaps the CK
elevation is simply a response to exercise, a stimulus that affects
the muscles on a daily basis in Duchenne patients.

DR. SCHWARTZ: I would like to make a point about the CK assays.
It seems that everyone measures the enzymatic activity but never
looks at the isoenzyme content of the protein. A study by Kuby
showed that in the Duchenne type patients there is an increase in
the mixed tissue (MB) form of CK whereas most normals have the
striated muscle form (13). I wonder why in all these assays the
isoenzyme is never measured?

DR. BROOKE: It has been measured repeatedly in a large series
of patients. The findings are being assembled and we hope to have
a report soon. There are, however, many previous studies of CK
isozymes in Duchenne dystrophy which don't reveal anything very
startling.

DR. SCHWARTZ: I would like to ask another question. In a
NATURE article, hexokinase isozymes are different in Duchenne
type patients (14). Has anyone followed up on those types of
experiments?

DR. BROOKE: I don't know the answer.

DR. HAUSCHKA: I would like to comment on Dr. Schwartz' statement
about the MB isoenzyme. We have studied CK isoenzyme in culture.
We find when myoblasts first differentiate they go from a situation
in which the only isozyme is the BB form, an increase in the MB
form follows and then finally the MM form appears. If one were to
see an elevation of the MB isoenzyme in the serum, it could reflect
new formation of myotubes as a result of the extensive muscle regeneration in these dystrophic cases. So it could be a useful thing
to check out consistently.

DR. BROOKE: That correlates exactly with what you see in the
muscle pathology. There are numerous fetal or undifferentiated
fibers in the Duchenne patient and I think that may correlate with
the levels of MB isozyme in the serum.

DR. LATT: Has anyone tried to compare the total increase in
activity measured in Duchenne with the amount of tissue that would
be required to produce it?

DR. BROOKE: We measured urinary creatinine, which is supposed
to mirror the muscle bulk: the more muscle, the more creatinine.
A plot of urinary creatinine as a function of age is linear as is
the plot of CK versus age. CK drops with age in Duchenne, perhaps
because there is less muscle and, as the disease progresses, urinary
creatine increases. With regard to CK efflux from muscle, I don't
know.

DR. LATT: What I was asking is can you compute the total
number of grams of tissue required to account for the total amount
of enzyme activity increase you get? If you take the total creatine
phosphokinase activity increment in Duchenne and you were to compute
how much tissue you need to account for that and it turns out to be
a hundred kilograms, you know all your assumptions were wrong. That
is one of the controls that is always necessary when you want to
interpret these enzyme activities as arising from this or that tissue;
at least are they internally consistent? Has anyone done it?

DR. BROOKE: I don't know.

DR. PATERSON: Has anyone ever measured the specific activity
of creatine phosphokinase during these bursts?

DR. BOYER: Hasn't this been measured immunologically?

DR. EPSTEIN: What we are raising here is that the CK activity
measurement seems to be much too loose a test. In fact, a quantitative immunological assay would be more reliable. At Baylor just
a gross crude activity is measured.

DR. STROHMAN: When you see increases in CK, is it coming
from the new fibers that are forming or from the old fibers?

DR. BROOKE: What we know about the morphology suggests that
the membrane in all the fibers is leaking. Tissue culture work
has not shown any difference between Duchenne and normal muscle
cells, possibly because muscle cells in culture are not fully
differentiated. Before differences will become apparent it may
be necessary to achieve maturity in muscle cell cultures.

DR. STROHMAN: That should be emphasized because it will
probably come up again and again. When you culture adult muscle,
whether it is dystrophic or normal, what is expressed in culture
is the embryonic state for the most part. Where you have clear
cases of embryonic to adult transitions, you never get the adult

form expressed in culture. What you tend to get is the regression back to the embryonic form. You have got to find a way to drive the cell so that it begins to express the pathological case.

DR. EPSTEIN: With respect to muscle cell phenotype in culture, some gene families will be expressed as the adult or as mixtures of the adult and embryonic forms whereas other gene families will be represented by only the embryonic forms.

DR. HAUSCHKA: In the case of the CK isoenzymes, if you call the MM form the adult form, you get that in all tissue cultures that differentiate just to the myotube level.

DR. NADAL-GINARD: It is important to establish, if possible, a correlation between the isoenzyme pattern in serum and in different muscles.

DR. HAUSCHKA: In muscle there is a sampling problem because the biopsies come from one muscle whereas the CK in the serum is from everywhere. Unless you are lucky there may be no correlation.

DR. NADAL-GINARD: That is my point about making the correlation between muscle histology and isoenzyme patterns in the CK of the biopsy and of the serum. This may permit us to find whether the serum CK is coming from degenerating muscle (MM) regeneration muscle (MB) or from everywhere (BB) or a mixture.

DR. BROOKE: I would caution the geneticists to make sure of the diagnosis in the patients whose X chromosomes you study. I can give you a set of criteria which you can use to evaluate whether somebody has Duchenne or not.

Criteria for Diagnosis of Duchenne Dystrophy

DR. EPSTEIN: Would your criteria prove to be generally acceptable to other neurologists?

DR. BROOKE: When we set up these criteria, we wrote to all of the people we knew who were interested in muscle disease and asked them to review the criteria for us. We had minor modifications but no significant disagreement among the group as to the criteria for the diagnosis of Duchenne dystrophy. The criteria that we laid down have been published in MUSCLE AND NERVE and are listed in Table 1 (15).

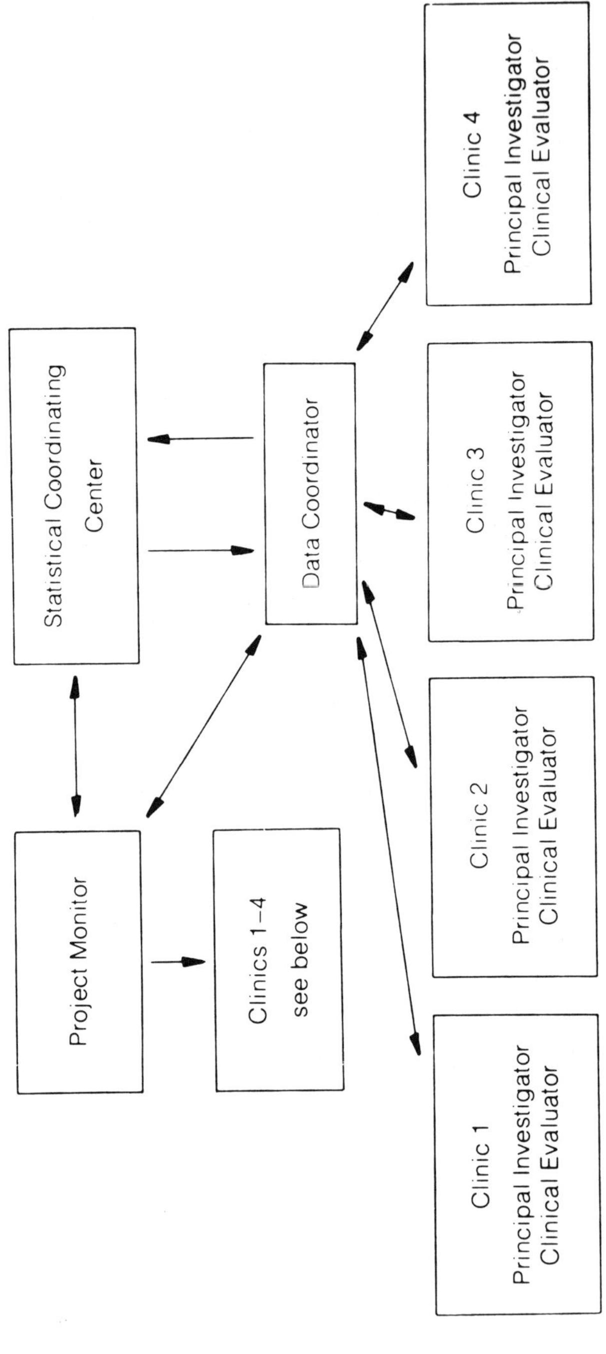

Table 1: The organizational arrangement of the multiclinic trial for Duchenne muscular dystrophy.

You will notice that an X-linked recessive family history is
not required for inclusion in the study. We decided to take the
child who fulfilled the clinical criteria for Duchenne and then
find out the family history rather than making it part of the
diagnostic criteria.

DR. SINISCALCO: You were very wise in deciding whether the
Duchenne clinically is a Duchenne on clinical grounds, thereby
avoiding the bias of a pedigree. You did introduce a bias, however,
by restricting the diagnosis to male patients. If you had maintained full objectivity you might have had to include some females.
If you then discovered that all of them had some X chromosomal
structural abnormality you would surely have concluded that this
must mean something.

DR. BROOKE: I would say categorically that what we now call
congenital muscular dystrophy in girls, apart from those with
Turner's syndrome is quite distinct from Duchenne. I am not sure
about the girls with an abnormality on the short arm of the X. I
think most of the clinicians even those who see a lot of muscle
patients will tell you they have never seen a convincing case of
Duchenne dystrophy in a girl.

DR. HECHT: One should always keep in mind that with very few
exceptions, the normal X is inactive in all cells of patients who
have the reciprocal X autosome translocation going through the
short arm of the X. The other possibility in _de novo_ heterozygotes,
which half of your females will be, is that they have the normal
allele in the structurally normal X. The location of these rearrangements may occur in other places that do not become apparent.
Even if clinically the cases fit the criteria of Duchenne, an
alternate explanation should be kept in mind until the diagnosis
is nailed down by another method.

DR. FRANCKE: There are now seven isolated cases of females
with so-called Duchenne muscular
dystrophy and X-autosome translocation.
Duchenne Dystrophy   The reported cases in females were
in the Female   clinically consistent with Dr. Brooke's
criteria except for being female.

DR. WILLIAMSON: Was muscle histochemistry done?

DR. FRANCKE: Muscle biopsy was done and examined histologically in all cases and histochemically only in one. EMG and CK were done. But the majority of those cases have only been published in abstract form.

DR. WILLIAMSON: Dr. Brooke, have you or your collaborators ever seen any of these seven cases?

DR. BROOKE: No.

DR. EPSTEIN: Is it conceivable that females are automatically genetically modified for Duchenne and therefore one should not use Brooke's criteria to reject them?

DR. UTA FRANCKE: The experience with Lesch-Nyhan (LN) disease may be considered in this context because it is also an X-linked disorder that is lethal in the male. Lesch-Nyhan Disease   No LN male is known to have ever produced a child. The patients have a severe neurological handicap in addition to mental retardation (16). The disease is not as common as Duchenne muscular dystrophy, but there is a biochemical marker for it: deficiency of the enzyme hypoxanthine-guanine phosphoribosyltransferase (HPRT) (17). The gene for HPRT maps on the X chromosome towards the distal end of the long arm. Genetic heterogeneity does not present a problem because demonstration of the biochemical defect proves the diagnosis.

Furthermore, based on the known biochemical defect and on X inactivation in females, reliable carrier detection tests have been developed. These include assaying HPRT activity in fibroblast clones (18). Alternatively, one may incubate a mixture of skin fibroblasts in selective medium containing either HAT (hypoxanthine aminopterin thymidine) or 8-azaguanine medium (19). The third and most rapid test has been developed by Stanley Gartler and his colleagues (20-21). It involves hair roots. A single hair root forms from about three precursor cells. When the precursor cells are set aside to form an individual hair root in the 11th week of gestation, there has already been coherent clonal growth of the scalp epithelium. This provides the basis for the finding that in mosaic females the majority of hair roots are of one or the other type. Because of differences in absolute HPRT activities between individual hair roots, the autosomal enzyme adenine phosphoribosyltransferase (APRT) is measured in the same sample (20) or in the same gel after electrophoretic separation of HPRT and APRT present in a single hair root (21). When you pluck individual hair roots from different areas of a woman's scalp and assay them for HPRT and APRT activities, you will find a significant number that completely lack HPRT in the presence of APRT if the woman is a Lesch-Nyhan heterozygote (20). With the combination of these

three tests, we have a reliable method of identifying carriers.
When these tests were applied in different laboratories to the
same females, excellent concordance was obtained (22).

Application of carrier detection tests (CDT) to mothers of
LN patients provides a direct means of testing Haldane's hypothesis.
The data indicate that much fewer than the 1/3 predicted are apparently
new mutations; the majority of LN mothers appear to be heterozygotes.

In 1975 I compiled all the data that were available in the
laboratories of Drs. Nyhan, Seegmiller, Gartler, Migeon and DeMars.
Of a total of 47 families, 27 had a single affected male (simplex
families) and 20 had more than one affected male (multiplex families)
(22). Additional studies were done in Dr. Nyhan's laboratory
between 1975 and 1980 on another 27 families with a single affected
and eight with multiple affected males (Table 2) (23).

Table 2
LN Families Studied with Carrier Detection Tests

Classification	OLD (1975)	NEW (75-80)	TOTAL
Simplex	27	27	54
Multiplex	20	8	28
TOTAL	47	35	82

Table 3 summarizes the results of the carrier detection tests
in both sets of families.

Table 3
Heterozygous Mothers (M) and Maternal
Grandmothers (MG) in Simplex Families

Test Result	OLD	NEW	TOTAL
M +/+	4	7	11
M +/-, MG +/+	5	5	10
M +/-, MG +/-	9	1	10
M +/-, MG not tested	9	14	23
TOTAL	27	27	54

In the previous studies there were only four mothers who were not
carriers. In the new series there are seven more that are not
carriers. Eleven instances of new mutation among 54 simplex families
is still below the 1/3 expected. We also looked at the maternal grandmothers in order to get at the question of how many mothers themselves were new mutations, and there were a total of ten. But in the
new series, there is a large number of maternal grandmothers who have
not yet been studied. In the ten families where mothers and maternal

grandmothers were heterozygotes the mutation must have occurred in an earlier generation.

Table 4 lists the ages of the maternal grandparents in those cases where the mothers were carriers but the maternal grandmothers were not. The +/- mothers must have received the mutant gene from either the father or the mother. In the first series there were quite a few older fathers, so we tentatively concluded that there could be a paternal age effect on the occurrence of new mutants for LN in heterozygous females (22). In the new series, however, only one grandfather was of advanced age. The average mean paternal age of 33.5 is somewhat higher than the mean paternal age for that particular generation in the United States, which is 29 years.

Table 4
Heterozygous Mother is a New Mutant
Her Parents' Ages at Her Birth

	Father	Mother
Old	27	24
	35	35
	40	31
	38	32
	40	39
New	21	20
	27	21
	40	43
Mean	33.5	29.4

We also examined the incidence of carriers among female relatives, maternal aunts and sisters of affected boys. In the first series, there was a vast excess of carrier females among the patients' sisters who were daughters of known heterozygotes, and when the grandmothers had been shown to be heterozygotes then the maternal aunts were also included (Table 5).

Table 5
Carrier Detection Tests on Daughters
of Proven Heterozygotes

Test	Simplex		Multiplex		Total
Result	OLD	NEW	OLD	NEW	
+/-	27	12	34	11	84
+/+	9	8	21	6	44

The new studies still show a slight excess of heterozygous females.
It is not clear whether this indicates some kind of heterozygote
advantage. If there were an increased segregation ratio of the
LN gene, which means a preferred fertilization of an oocyte that
carries the mutant gene, one should see an excess of LN males as
compared to unaffected brothers, unless there were selection against
mutant males in utero for which there is no evidence. This problem
is currently being studied.

In summary, the data on LN families, as recently updated and
statistically analyzed (23), are consistent with an increased mutation
rate in males over that in females. Alternative explanations of
the data have been considered and are felt to be less likely (23).

DR. DAVID HOUSMAN: To compare the diagnostic criteria for
Huntington's disease with those of Duchenne dystrophy, I think a
positive family history is crucial
to making the diagnosis of Huntington's.
From the genetic point of view, the
only cases that are important to
study are those in which a significant
pedigree exists in which X-linked inheritance is clearly demonstrated.
From the point of view of the clinician, information from those
cases with a clear cut family history will enable you later to
categorize those cases which do not have positive family history.
I think that evidence for X-linked is the key issue, whether we
call the disease Duchenne or not.

DR. BOYER: I continue to urge the study of sib pairs, noting
the heterogeneity between pairs. If one of those falls outside of
the criteria that you have set and the other is bang on, I would
suggest that the one you threw out really does have Duchenne and
is modified. And that, of course, would be a matter of great
interest since it would lead one to a locus that would modify the
disease.

DR. EPSTEIN: Dr. Housman, please be more specific as to what
you mean by a clearcut family history. It is conceivable that
many people might call a sibship with two males having Dr. Brooke's

criteria as evidence for family history and yet one could immediately
think of alternative mechanisms.

DR. HOUSMAN: The larger the pedigree and the more affected
individuals in different generations, the more informative the
pedigree will be.

DR. WILLIAMSON: As the family size increases, you get a geometric increase in the certainty of your linkage prediction. Thus,
the family with seven or eight unequivocal cases plus unaffecteds
and carriers is more useful than 20 families, in each of which
there are only two or three affected children.

DR. BROOKE: We have about 10-12 families in which there are
multiple members involved, as many as five boys in the same family
with about 10 uncles who are also living. We will separately analyze,
from the point of view of the clinical course, the data on all of
the boys who have brothers with the same illness. This will tell
us whether the people in the same family do, indeed, have the same
clinical course. We have made over 200 measurements on each child
at each clinic visit. The data are not ready yet but my impression
is that there is variation in the IQ. I think the other parameters
don't vary very much.

DR. WOLF: Dr. Brooke, roughly what percentage of your patients
are sporadic cases?

DR. BROOKE: About 30 percent.

DR. HOUSMAN: In the Huntington's disease roster in which
families from all over the country are being analyzed, some of the
pedigrees are connecting up and that increases the power of the
analysis enormously. I think it will be very important if such a
roster could be developed for Duchenne dystrophy.

DR. CONNEALLY: Also one should begin storing DNA from these
families as we are doing with the Huntington's.

DR. NADAL-GINARD: I think the goal is to find biochemical
markers for Duchenne muscular dystrophy. Since we define Duchenne
as an X-linked muscular dystrophy,
Defining the Goals   we should study X-linked muscular
dystrophies and worry about new
sporadic cases later. Dr. Brooke outlined the major diagnostic
criteria for Duchenne dystrophy. When Dr. Francke mentioned the
female cases, one of the minor criteria had to be brought in to
be able to distinguish whether they were Duchenne's or not. I
think we should concentrate first on the notion that Duchenne has
a family history that fits an X-linked pattern.

DR. CASKEY: I disagree. I think we ought to try to embrace
as much of the data on Duchenne as we possibly can in considering
experimental approaches. If we just concern ourselves with linkage,
we will miss a good opportunity to get a handle on the gene. We
all agree that if we pick up a linkage association, we may be a
long way from the gene we are primarily interested in. We should·
therefore give consideration not only to linkage analysis in
families with multiple affecteds but also to rare occurrence which
may, in fact, target us more precisely to altered DNA sequences.

DR. NADAL-GINARD: I don't disagree with what you said, but
we cannot use a shotgun approach. We know that on the X chromosome
there is a gene that produces muscular dystrophy. I think that we
all agree on that and we are going to have our hands full to find
the gene. It would be better to concentrate on something that we
can identify on the genome.

DR. EPSTEIN: Bernardo, although operationally what you say is
laudable, there are some people who would argue that we don't really
know that there is a gene. One of the hypotheses that has been
raised is that there are multiple loci. That may not be a good
hypothesis but it is an hypothesis. The other possibility that
mutations may commonly be caused by insertion or deletion of
genetic elements as shown by recent work with drosophilia and yeast
fits what Dr. Caskey suggests in view of the very high sporadic
mutation frequency in Duchenne. It may be that an insertion at any
place over a fairly large region of the X chromosome causes Duchenne
and that there isn't a specific Duchenne gene.

DR. NADAL-GINARD: I am not talking about the specific gene.
In these sporadic cases you don't know whether they are linked or
not. So you don't know in what chromosome the problem is. I
am suggesting that we concentrate on the X chromosome as identified
by the pattern of a pedigree.

DR. WILLIAMSON: Some people will concentrate on some things
and some on others. We should bear in mind that there are two
types of heterogeneity which must be distinguished to avoid
confusion later. The heterogeneity in the β-globin thalassemias
reflects a large number of different molecular events occurring
at the same locus, all leading to approximately the same clinical
syndrome. Dr. Boyer was referring to the other type of heterogeneity where a genetic event at one locus is modified by something
happening at another locus. Thus, there are different forms of
genetic heterogeneity with very different implications from the
point of view of linkage analysis. The first type will show linkage
to a single locus and will actually be indistinguishable from a
linked marker. The second will show vast differences in linkage
and therefore will have to be analyzed in a different way.

DR. SINISCALCO: Since there are females with alleged muscular dystrophy in whom there is a translocation on the short arm of the X, the same lesion should occasionally occur in males. We studied, with the help of the Muscular Dystrophy Association, 26 families with muscular dystrophy in Sardinia, collected by the Institute of Neurology of the University of Sasseri. In none of these cases or their mothers did we find a single one who had an abnormal X at the level of Xp21. In case someone might think this is a very common type of event, it is not.

CHAPTER 2

CLONING FOR AN UNKNOWN GENE

DR. NORMAN DAVIDSON: In order for a gene to be isolated and
characterized by recombinant DNA methods, characteristics or phenotypes must be known. Thus, while there are no methods of cloning
for an unknown gene, there are many published examples of methods
of cloning for a gene coding for a protein that is abundantly, or
moderately abundantly, expressed in a given tissue type.

A cDNA library is constructed as a first step. To do this,
messenger RNAs (mRNAs) are isolated from the cells of the higher
organism under study. DNA copies (cDNA) of these mRNAs are synthesized and then inserted into circular DNA plasmid molecules.
These plasmids can infect and transform E. coli bacterial cells.
A cDNA library is a collection of bacterial clones that contain
such cDNA molecules.

One generally starts by constructing a cDNA library, that is,
a set of bacterial clones. Each cell in any one colony contains
circular plasmids (DNA molecules) which can replicate in E. coli
and which include an insert which is a cDNA copy of one member of
the small mRNA population. The colony containing the gene of
interest is selected by hybridizing its DNA with a cDNA probe
(i.e., a radiolabeled cDNA molecule from an mRNA) which has been
highly enriched (perhaps by gel electrophoresis or sedimentation)
for the gene of interest. Of the several or many clones so selected,
one or several may correspond to the gene of interest. These can
be positively identified by a variety of methods, depending on what
is known about the particular protein. For example, if all or part
of the protein sequence is known, this sequence can be compared with

that deduced by determining the DNA sequence. An alternative procedure of general applicability is to use the cloned DNA to extract
complementary RNA which is then subjected to in vitro translation
(23). The in vitro translation product can be identified by two-
dimensional gel electrophoresis, or by specific antibody precipitation if an antibody is available. In some cases, by carrying out
the translation in the Xenopus oocyte, one can obtain sufficient
protein to demonstrate enzymatic or biological activity.

A particular gene may be expressed in cells after an appropriate
stimulation, for example by hormones, or at one particular stage of
                                 development, and may not be expressed
Isolation of Specific Genes   without induction or at a different
                                 stage of development. In these cases,
one chooses for study only the subset of clones which hybridize to
probes from the expressing cells and do not hybridize to probes from
non-expressing cells. This is a very useful general method for
restricting the number of clones that need to be studied. Many of
these genes are of particular scientific or medical interest.

There are many such examples including the now classical cases of
the isolation of globin genes (24), immunoglobulin genes (25), and
actin genes (26). The genes coding for the DOPA decarboxylase enzyme
and for the larval cuticle proteins of Drosophila were isolated by
related approaches, even though these proteins are expressed at
levels of 0.1 percent or less of total cell protein (27,28).

When the protein sequence is known, one may synthesize a set of
oligonucleotides (usually of length 11-20) coding with a minimal
degree of degeneracy for a short oligopeptide within the protein.
The mixture of oligonucleotides can be labeled and used as a
hybridization probe to select those cDNA colonies coding for the
protein of interest. This method is effective even when the protein
is expressed at a level of 0.01 percent of total cell protein and
perhaps even less. An example of this is the isolation of the
β_2-microglobulin gene (29).

The more interesting methods that I wish to consider are those
that apply to genes for which the protein is not well characterized
or is insufficiently abundant but for which the gene can be recognized by its phenotype when reintroduced into mammalian cell. It
is known that when cells in culture are exposed to exogenous DNA
present as a co-precipitate with calcium phosphate, a small fraction
of the cells take up the exogenous DNA by stable incorporation into
the host chromosome. All progeny cells from any one such transfected
cell will contain the exogenous DNA. In some cases, the gene or
genes on this foreign DNA are expressed at a sufficient level to be
detected by their phenotype or by hybridization methods. This general
subject has been reviewed recently (30).

The great value of this procedure lies in its ability to isolate genes with very low levels of expression. One example is illustrated in Figure 6 (31). A murine cell line is available which is defective for APRT.

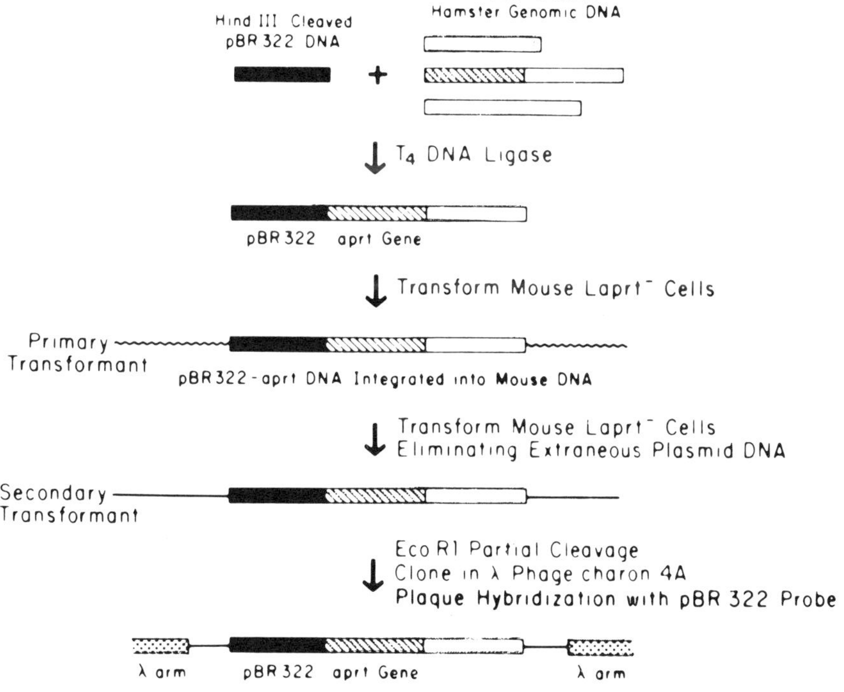

Figure 6: Procedure for the isolation of the hamster APRT⁻ gene.

Cells which are APRT⁻ do not grow in media containing appropriate
concentrations of azaserine and adenine. Cells which are APRT⁺ do.
When cells which are APRT⁻ are treated with DNA from a cell which
is APRT⁺, a small fraction of these cells acquire the exogenous
APRT gene and can be selected by growth in the selective medium.
It should be recognized that only about one DNA molecule out of 10^5
of the exogenous DNA contains the APRT⁺ gene. In the experiment
described in the figure, we know that the small fraction of murine
cells which have become APRT⁺ contain the hamster APRT gene. How
can we recognize this exogenous hamster DNA against the background
of 10^5 parts of murine DNA, in view of the fact that there are
many cross-hybridizing sequences between hamster and murine DNAs?
One way of doing this is illustrated in Figure 6. Each molecule
of hamster DNA is first attached to some DNA which does not have
a sequence related either to hamster or murine DNA. In the case
in Figure 6, that DNA was the E. coli plasmid pBR322. In the
actual procedure those cells which have taken up the DNA molecule
containing the hamster APRT gene will usually also take up about
100-1,000 additional hamster DNA molecules. Therefore, DNA from
the primary transformants is again transferred to recipient APRT⁻
cells to eliminate extraneous plasmid and hamster DNA and to
construct secondary APRT⁺ transformants containing only mouse DNA
and the hamster APRT gene which is linked to a single plasmid
sequence. These plasmid sequences can then serve as a physical
marker for the APRT gene.

DNA from the secondary transformants was used to construct a
library of recombinant bacteriophage in a suitable λ cloning
vector. This recombinant library was screened with highly radioactive pBR322 DNA as a hybridization probe, and those recombinants
containing plasmid sequences were tested for the presence of the
APRT gene. This method indeed resulted in the isolation and
subsequent characterization of that gene (31).

The method as described above requires the attachment of the
exogenous DNA to a prokaryotic segment, such as pBR322. The
presence of characteristic (that is species specific) short, highly
repeated sequences in a genome may make this latter step unnecessary.
For example, human DNA contains a sequence of length approximately
300 bp which is present approximately 6×10^5 times and appears to
be more or less randomly distributed in the genome (32). Other
highly repeated human specific sequences are also present in the
human genome. The general situation is depicted in Figure 7. It
is therefore probable that at least one such sequence would be
present in over 95 percent of all clones containing 15-20 kb inserts
of human DNA (for example, an insert containing the human APRT⁺
gene if we wished to isolate this gene by the method described
above). These inserts will hybridize with an Alu-specific probe.

INTERSPERSED REPEAT SEQUENCES CAN BE USED AS SPECIFIC PROBES
FOR HUMAN DNA IN A BACKGROUND OF DNA FROM OTHER SPECIES

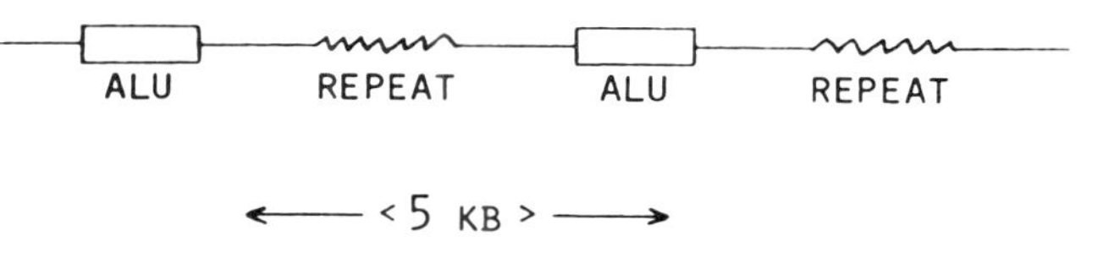

CAN USE TOTAL NICK TRANSLATED HUMAN DNA AS A HYBRIDIZATION PROBE
(ONLY HIGHLY REPEATED SEQUENCES HYBRIDIZE)

OR CAN USE A CLONED ALU FRAGMENT

USE THESE PROBES TO ISOLATE CLONED GENES FROM A PARTICULAR
CHROMOSOME (OR FRAGMENT) FROM A LIBRARY OF DNA FROM A
HYBRID CELL

Figure 7: Schematic representation of the distribution of highly repeated short interspersed repeat sequences in the human genome. It is estimated that the average spacing between Alu sequences is 5-6 kb, and that there are an approximately equal number of other highly repeated sequences in the genome.

An Alu-specific probe can thus be used to recognize a small fraction
of human DNA clones in a murine or other background. Gusella et al
(33) have shown how to use this method that selects from the cloned
library of DNA from a cell line containing human chromosome 11 in a
Chinese hamster background, those clones with inserts from human
chromosomes. In this instance, the authors used hybridization with
all of the highly repeated human sequences, not simply the Alu
sequence, to identify human inserts.

A possible application of this method for the isolation of
specific oncogenes from human tumor cell lines has been described
by Murray et al (34). In this case, the selectable phenotype
is the ability to form a transformed focus. This phenotype is
acquired by certain murine fibroblasts after transfection of human
DNA from colon or bladder carcinoma cell lines. It should be noted
that the work published so far shows that a specific fragment of
human DNA containing Alu sequences is associated with the transforming gene. It may be presumed that this procedure will enable
the authors to actually clone and isolate the transformation gene.

Genes without a selectable phenotype can be introduced to cells
in culture along with a selectable gene by the procedure of co-
transfection (30). At present this procedure is useful for the
study of an expression of an already cloned gene in an exogenous
background. An example from the work of Goodenow et al (35) is
illustrated in Figure 8. A cloned histocompatibility gene was
cotransfected into mouse L TK$^-$ (thymidine kinase deficient) cells by
cotransfection with the Herpes virus TK gene. (TK$^-$ cells do not grow in
HAT medium, TK$^+$ cells do; the selection is excellent). A reasonable
fraction of the transformed TK$^+$ cells also takes up the H-2
(mouse major histocompatibility locus) gene. The cloned gene was
known by hybridization criteria to be related in sequence to the H-2
class I genes. However, it is very difficult by DNA sequencing and
other recombinant DNA methods, to identify a specific H-2 class I
DNA gene because of the tremendous polymorphism of the H-2 locus,
because there is good cross-hybridization between different H-2
class I genes, and because all the protein sequences are not yet known
However, the gene can be identified if it is expressed in an
exogenous background because of the exquisite specificity of the
serological reagents available for H-2 gene products. In the present
instance, the gene was shown to code for an H-2L^d. The gene was
derived from a BALB/c mouse which is of the H-2^d haplotype, and is
expressed in C3H mouse fibroblasts (H-2^k haplotype). This general
approach will be an elegant and powerful method for identifying
individual members of a complex multigene family that can be distinguished by serological methods.

In principle, it should be possible to start with a library
of DNA from an organism which expresses a particular cell surface
antigen, transfect that DNA into mammalian cells which are not

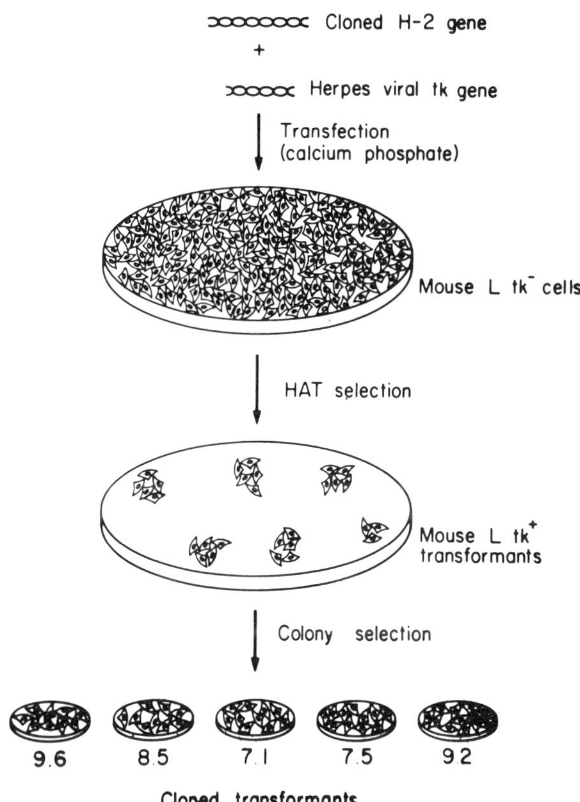

Figure 8: Procedure for identifying a cloned H-2 gene by serological identification of the gene product after transfection into a suitable mammalian cell.

expressing that antigen, and then using an immunoselection procedure (possibly by flow cytometry), to select those cells into
which the gene of interest has been transfected. At present, the
efficiency of immunoselection and the throughput in flow cytometric
procedures, and the efficiency of DNA mediated transfection are
not yet high enough to make this proposal practicable. However,
we anticipate that these technical problems will be solved and it
will be possible in the near future to select for any gene which
can be recognized as a cell surface antigen by immunological
procedures. Alternate procedures for identifying genes that are
not cell surface antigens but which have a phenotype, for example,
an enzymatic activity which can be recognized histochemically, can
be conceived of and will probably become practicable soon. Thus,
in principle, a gene which can be recognized by its phenotype in
cell culture should be clonable.

If research on muscular dystrophy reveals a cell biological
phenotype expressed by cells carrying the defective gene or genes
that cause the disease of muscular dystrophy, the approaches described above may be useful for isolating and characterizing the
gene.

We have excluded from our discussion the proposed method of
mapping genes with an observable phenotype in a whole organism by
the method of pedigree analysis, which correlates segregation of
the trait of interest with mapped restriction site polymorphisms
of cloned DNA inserts (36). This method is discussed elsewhere
in this volume. If a specific DNA insert associated with Duchenne
muscular dystrophy is isolated by this method, its cell biological
phenotype can be studied by the methods of DNA transfection.

DR. GUSELLA: A group at the University of Toronto have cloned
                              a leukemia associated cell surface
Promoting Gene Expression   antigen by transformation into a
                              mammalian cell but not looking for
stable integration just for transient expression (37). They have
just worked by cutting down the size of their library in subclone
selection and cloned several copies of this genomic gene.

DR. DAVIDSON: Let me say in that connection that Dr. Gusella
has mentioned something important. Namely there are techniques for
transient expression which give a much higher fraction of total
cells expressing for several days as distinct from the stable integration. These methods may turn out to be very useful.

DR. PATERSON: What kind of background do you put that gene in
if you want stable transformants. Let's say you take a gene that
has never been expressed in those cells, and you are trying to
immortalize it the way you described. What if the environment is
such that the gene would never be expressed?

DR. DAVIDSON: I think we are learning enough about sticking
SV40 or retrovirus promoters onto random chunks of DNA so that we
can up the expression of any gene to a level where it can be detected
even if it is normally expressed at a very low level.

DR. NADAL-GINARD: The problem is you have to use the SV40
promoters or retrovirus promoters to select for the gene. You may
have a serious problem to know where to put the promoter.

DR. DAVIDSON: There is evidence that gene expression is en-
hanced by the SV40 enhancer sequence anywhere in the vicinity.

DR. HAUSCHKA: To respond to Dr. Paterson's comment regarding
the possible importance of analyzing the expression of cloned genes
in an appropriate host cell, Dr. Gary Merrill, an MDA Fellow working
in my lab, has derived an immortal line of thymidine kinase-minus
mouse myoblasts which he has now successfully transfected with cloned
Herpes virus TK. One strategy, then, would be to attempt to co-
transfect these cells with various cloned muscle genes. Such experiments might eventually lead to co-transfection with the human
muscular dystrophy gene.

DR. DAVIDSON: So little has been done actually. I think I
have covered everything which has been done in the line of isolating
genes with some sort of a selectable phenotype or immunofluorescence
selection. It seems to me that important questions in the discussion is what is known about the cell biology of muscular dystrophy;
what sort of changes are there on the cell surface? How clear is
the evidence whether the primary defect is in the muscle cell or in
the nerve cell through the influence of some trophic factor? When
will our knowledge of cell biology be far enough advanced so that
one can begin to answer the questions? I hope the people who know
something about that will contribute as the afternoon goes on.

DR. NADAL-GINARD: In order to do the transformation, you don't
need to have TK⁻ cells even though the TK⁻ is a very good gene to do
it with. But you can use dominant markers. You don't need to have
the mutant.

DR. DAVIDSON: That is correct. There are several dominant
markers that one can use for transformation into any cell but
one must employ this low frequency of 50-100 per tissue culture
dish which is not enough for screening a library.

DR. HAUSCHKA: One technical hurdle to obtaining transfectants
with myoblasts is that when one plates high density for transfection
the cells become post-mitotic and differentiate before transfected
clones can be isolated. This occurs because high density cultures
rapidly deplete growth factors from their environment; then, when
mitogens fall below threshold levels for 2-3 hours, myoblasts commit

irreversibly to terminal differentiation. Thus, even if a cell
had been transfected it would respond to environmental cues and
differentiate rather than form a recognizable colony. When we
finally realized this was occurring we discovered that _more_ transfectants could be isolated by _reducing_ rather than _increasing_ the
densities of myoblasts treated with DNA.

DR. HOUSMAN: Isn't calcium a signal to these cells to differentiate?

DR. HAUSCHKA: No. Myoblasts require calcium for fusion; but
many investigators have shown that muscle differentiation occurs
in the absence of fusion and at low calcium levels.

DR. HOUSMAN: So you are treating them with very high concentrations of calcium when you put the DNA on. Does that have any
influence on their fusion?

DR. HAUSCHKA: Probably not. The transfection protocol does
call for high calcium levels; but calcium is then reduced to normal
levels and the cells are grown for many generations prior to stimulating their differentiation by mitogen removal. At least in the
case of transfected mouse myoblasts, such cells appear to differentiate and fuse normally.

DR. NADAL-GINARD: In terms of what Dr. Hauschka just mentioned,
we have a cell line that is temperature sensitive for commitment.
It does not commit at 40°C even though it differentiates, but the
cell does not withdraw from the cell cycle. In fact, we had tested
those cells and they take up DNA quite well. So you do the transformation at 40°C but you can do it at any density that you want.
Then you switch them to 33°C and they differentiate quite well.

DR. EPSTEIN: There really isn't any good answer to Dr. Davidson's last question. Our knowledge
of cell biology of human muscle is
Achieving Full   primitive, in comparison to some of
Differentiation _in vitro_   the work that has been done in chick
and rat, in terms of differentiation, isoforms, etc. As best as I
know, no one has gotten a population of human muscle cells to all form
sacromeres and other myofibrillar apparatus. Maybe 20 percent of the
cells have filaments but not organized ones and maybe one percent of
the cells actually form sacromeres.

DR. WOLF: Dr. Hauschka, is it reasonable to guess that
innervation might be required in order to achieve full differentiation.
That is true of smooth muscle cells.

DR. HAUSCHKA: Yes, it is possible that innervation may be required to achieve full differentiation of muscle _in vitro_. Perhaps
Dr. Strohman could comment further on this possibility. In reply

to Dr. Epstein's comment, one of the great difficulties of studying
these questions with human muscle cells is that permanent myogenic
lines have yet to be achieved. There is some promise from the work
of Dr. Armand Miranda at Columbia that human myoblasts can be transformed with SV40; but it is not yet clear that the resulting cells
have retained their myogenic phenotype.

DR. PATERSON: At Cold Spring Harbor, Dr. Helen Blau presented
some preliminary data that looked very encouraging in terms of establishing human cell cultures, and she had some micrographs of
clearly striated human muscle in culture. I have seen Duchenne
tissue as well as the normal human culture material. It is a
tedious approach, but it seems to be working in her hands. People
should be aware of the fact that she is doing it.

DR. EPSTEIN: She reports a very high percentage of fusion and
that 80 percent of the myotubes are striated.

DR. BROOKE: Be careful when you talk to muscle histochemists
or muscle pathologists about differentiation because they are talking
about something else. They are not talking about the occurrence of
cross-striations which is not difficult to achieve in muscle cultures.
They are talking about cells with biochemical, physiological and
structural properties of the differentiated adult form cell and that
has not yet been achieved in muscle culture.

DR. STROHMAN: I think one has to be clear that the muscle cells
in culture are well developed. They are cross-striated, they are
physiologically contractile, but so far as I know they do not express
anything but the embryonic phenotype. That is if you culture from
adult dystrophic muscle, you will get in culture satellite cells that
will proliferate into myoblasts that fuse to form cross-striated
myotubes. They express embryonic kinds of myosin and other isoforms.
Muscle cells in culture do not go through the normal embryonic to
adult transition.

Muscle cells from DMD patients appear normal in culture, perhaps
because they express only a normal embryonic phenotype. We need to
create the cell biological conditions in culture which permit the
cells to mature beyond terminal differentiation. It is a whole other
kind of ballgame.

DR. WOLF: Has that been tried with nerve cells in the culture?

DR. STROHMAN: Yes. It looks like one can perturb the expression
of myosin light chains by co-culturing skeletal muscle with different
forms of nerve cell lines. But that is very preliminary.

DR. NADAL-GINARD: You said something that I think is very
important. Duchenne is probably a disease of the adult muscle. So
we have to have a way to make an adult muscle _in vitro_. It is very

clear, for example, that thyroxine plays a role in switching the
genes for myosin heavy chains in cardiac muscle. You can switch
from an embryonic-like to an adult type of gene just by thyroxine
and you can go back and forth many times. Probably it would be
interesting to check on whether or not thyroxine or any other hormone
or serum from adults can or cannot induce the switch _in vitro_.

DR. MARTONOSI: I wonder if someone could clarify for me what
kind of phenotypic features we are looking for. In this discussion,
adult and embryonic muscle and dystrophic muscle phenotypes have
been mentioned. I don't know how I would recognize any of these if
I would have them in tissue culture. Could you sum it up for me so
I could understand?

DR. STROHMAN: We don't know what to look for in terms of having
a dystrophic myofibril in a dystrophic muscle fiber in culture. In
comparing embryonic to adult muscle phenotypes we can look at peptide
maps of myosin heavy chain. We can clearly distinguish an embryonic
form of myosin heavy chain from an adult form of myosin heavy chain
or distinguish the myosin heavy chain of a fast or slow muscle.
There have been reports in the literature that a myosin heavy chain
peptide map is unique to dystrophic muscle. I would say that is
almost true. In the dystrophic condition, embryonic muscle fibers
grow up from satellite cells and make embryonic kinds of myosin
heavy chains. This may account for all of those dystrophic specific
peptides of the myosin heavy chain. There is the need to do single
fiber analysis using the biochemical technology we have available.
We need to ask what the adult fibers are doing in a dystrophic
muscle as opposed to what the embryonic fibers are doing. Single
fiber analysis using the most advanced biochemical technology could
tell us a lot.

DR. LATT: In the absence of a good phenotypic marker for the
Duchenne muscular dystrophy (DMD) gene, an alternative goal is the
acquisition of pieces of DNA on the X chromosome which can be linked
to the DMD locus. A number of investigators have employed various
strategies to isolate pieces of DNA from the X chromosome. For
example, Wolf, Migeon and associates (38) have constructed plasmids
containing DNA from cells containing supernumerary X chromosomes,
from which X-specific sequences have been identified. Schmeckpeper
et al (39) and Bruns et al (40) have employed human-rodent cells
retaining the human X, but few if any other human chromosomes from
which phage libraries serving as a source of X-specific inserts
have been constructed. Finally, flow sorting has been employed by
Davies, Williamson and associates (41) and by our lab (42,43) to
enrich for the X chromosome and hence construct phage libraries
from which X-specific sequences can be isolated with very high
efficiency.

The following features are relevant when considering flow
cytomeric enrichment of mammalian X chromosomes (Figure 9). First
of all, a typical <u>sorting</u> rate for a given chromosome is a few
hundred thousand per hour. This
assumes an <u>analysis</u> rate of perhaps
1000 metaphase chromosomes per second.
It is possible to enrich 10-40X for
the X chromosome, which may be present in one, two, or, in some cells,
as many as five copies per 40 (mouse) or $\geq$ 46 (human) chromosomes.
Since chromosome enrichment provided by flow sorting is not biased
by sequence repetition, it complements the somatic cell hybrid
cloning approach introduced by Gusella et al (44). In principle,
flow sorted chromosomes can serve as a source of chromosomal proteins, although analysis of such material has not yet been published.
Finally, Disteche et al (45), under appropriate circumstances,
showed the presence of a structurally abnormal X on an analytical
scale.

Flow Sorting of
X Chromosomes

1. SORT RATE = FEW x 10^5/HOUR

2. CAN ENRICH 10-40 X FOR SPECIFIC CHROMOSOME

3. SELECTION NOT BIASED BY SEQUENCE REPETITION

4. CHROMOSOMAL PROTEINS ACCESSIBLE

5. SOMETIMES CAN RESOLVE ACTIVE AND INACTIVE X

Figure 9: Some features of metaphase chromosome
flow sorting.

Chromosome flow sorting can serve for X-specific DNA isolation.
Dr. Christine Disteche in my lab in collaboration with Dr. Eva
Eicher from Bar Harbor used the mouse Cattanach translocation
chromosome which has a piece of mouse chromosome #7 inserted into
the X, making it the largest chromosome in the cell (46), to isolate
X-specific DNA. Late replicating chromosomal regions in such cells
could be highlighted by initiating culture in the presence of BrdU
finishing in the presence of thymidine and staining with 33258 Hoechst.
Figure 10 shows one such cell, with brightly fluorescing mouse
centromeres, which contain satellite DNA, a normal X chromosome,
and the Cattanach translocation chromosome X(7) which appears both
large and bright; a tempting target for isolation.

As a prelude to flow cytometric work, we collaborated with
Dr. A.V. Carrano at Lawrence Livermore Laboratory to obtain a DNA
based karyotype of the mouse using gallocyanine-chrome allum as the
stain (45). This analysis demonstrated that the X(7) had about
15 percent more DNA than the next largest mouse chromosome (#1).

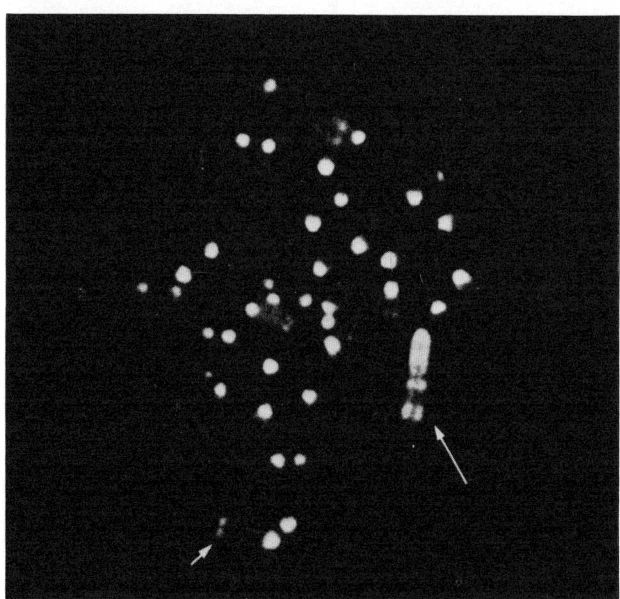

Figure 10: Late replication in a mouse embryo cell
carrying the mouse Cattanach translocation. The cell
was cultured in medium containing BrdU during the first
but not the last part of S phase. After staining with
33258 Hoechst, late replicating regions exhibit bright
fluorescence. The X(7) is indicated by the large
arrow.

One could then simulate what a flow histogram would look like
(Figure 11) to show that there should be a shoulder, containing the
X(7) chromosome, at the high fluorescence side of this histogram.
Dr. Disteche then established cell cultures from mouse embryos containing the X(7), and we carried out experiments with Dr. Carrano
to determine whether the theoretically expected resolution of X(7)
could be realized experimentally.

At Livermore, metaphase chromosomes isolated from mouse embryo
cells were stained with 33258 Hoechst and analyzed in a flow cyto-
meter equipped with a Spectraphysics Model 171 argon ion laser,
which can provide more than one half watt of stable illumination in
the near ultraviolet. Shown is a DNA flow histogram of such chromosomes (Figure 12). For reference, at the top, is a histogram from
normal female chromosomes. At the bottom is a histogram from a
male cell containing the X(7). The X(7) appears as a new peak of
high fluorescence, exceeding that of the largest autosomes by 25
percent (i.e. more than predicted), and clearly separated from the
normal X.

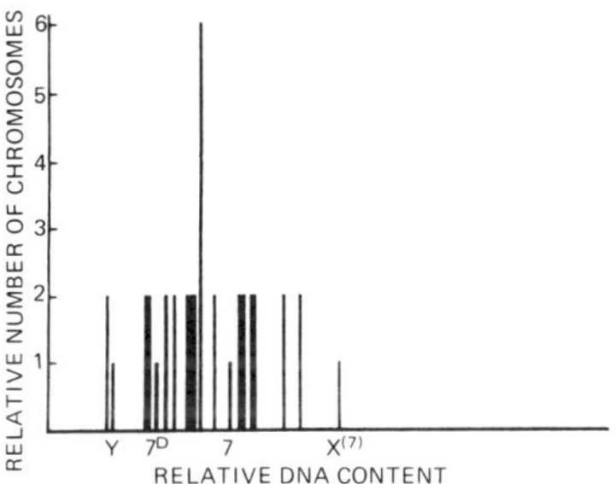

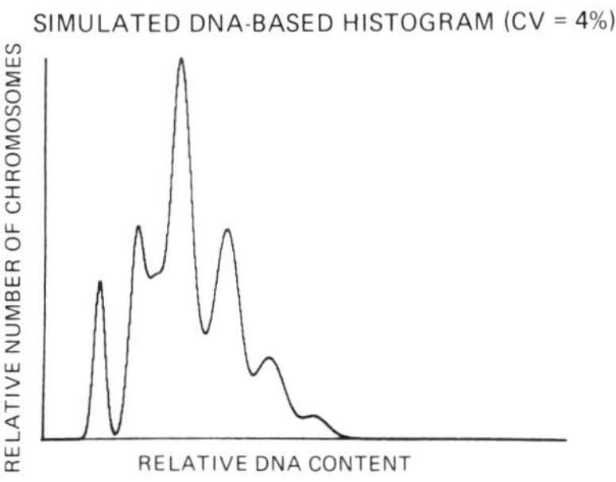

Figure 11: Simulated flow histogram of metaphase
chromosomes from a male mouse cell carrying the
Cattanach translocation. DNA content per chromosome
(top) was based on scanning cytophotometry (Disteche
et al 1981) (45). The simulated flow histogram
(bottom) assumed the fluorescence based on these
DNA contents was broadened by a coefficient of
variation (CV) of four percent.

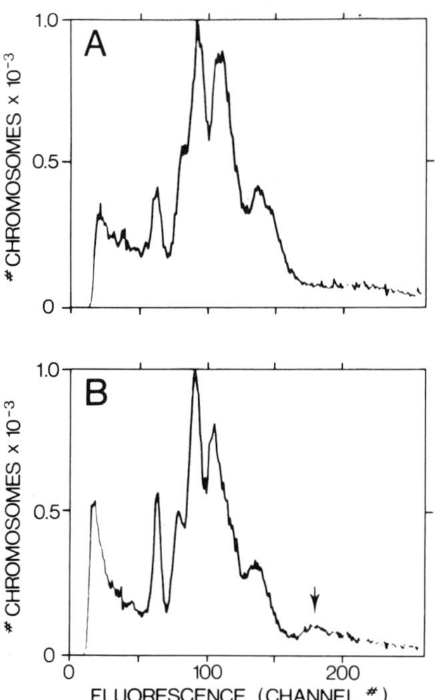

Figure 12:   33258 Hoechst fluorescence distribution of chromosomes: top, normal female mouse
cell line; bottom, a male derived cell line
carrying an early replicating X(7).

     Chromosomes flow sorted from the X(7) peak and stained with
quinacrine were more than 10 fold enriched for the X(7), to a total
content of 30-50 percent (Figure 13).

     Dr. Disteche then carried out flow sorting in our laboratory
on a preparative scale (42).   Several days of sorting were necessary
to obtain sufficient DNA for a phage library.   A few tenths of a
microgram of DNA were isolated from 2-3 million chromosomes.   The
object was to clone this DNA in relatively small pieces, many free
of repeated sequences that would thwart attempts to establish chromosome specificity.   The DNA was thus digested to completion with
EcoRI endonuclease and cloned in the vector λ gtwes, provided
by Dr. Stuart Orkin.   Dr. Disteche obtained one hundred thousand
plaque-forming units, in which the average size of a cloned insert
was 6-7 kb.   Most of the phages were recombinant.   The next
step was to plate out phage from the library and then screen these
phage for highly repeated sequences, first by the method of Benton
and Davis and then by Southern blots.   Repeat-negative phage were
then checked for X-specificity.

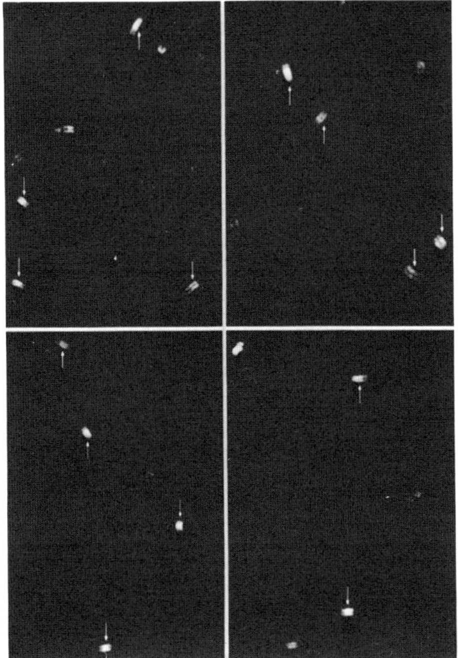

Figure 13: Examples of sorted chromosomes from the
high fluorescence peak of a cell line derived from
a male mouse carrying the Cattanach translocation.
The sorted chromosomes were spun onto slides, fixed,
and stained with quinacrine. Arrows indicate X(7)
chromosomes (Disteche et al, 1981) (45).

     Figure 14 shows a Benton-Davis screen, which provided first
evidence that we had obtained recombinant phage. Approximately
one third of the plaques were positive for repeated mouse DNA.
Plaques scored as negative were picked, amplified, and used as a
source of DNA which was further screened for repeated sequences
by Southern blots (Figure 15). Phage inserts putatively free of
repeated sequences were then used as probes of blots against
EcoRI digested DNA from a hybrid cell (obtained from Dr. Roseanne
Farber) that had the mouse X on a hamster background, DNA from the
parent hamster line, and DNA from male and female mice. Figure 16
shows blots, obtained by Dr. Lou Kunkel and Mr. Michael Eisenhard,
illustrating hybridization patterns of mouse X-specific and non
X-specific inserts. The former hybridized more intensely with
female mouse DNA than with male mouse DNA, and hybridized with
DNA from the mouse-hamster hybrid but not with the hamster DNA.
In contrast, the non-X-specific probe hybridized equally with male
and female DNA and exhibited a band of different size, with both
hybrid and hamster DNA samples.

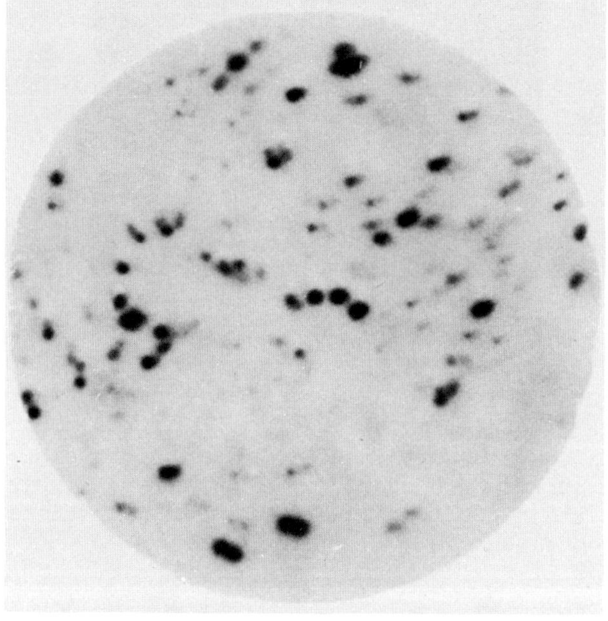

Figure 14: Benton-Davis (1977) (51) screening for
phage containing repeated mouse DNA sequences. This
X-ray film exposure is from a filter imprint of
plaques from a phage library enriched for X(7)
sequences that was probed with ^{32}P-labeled nick
translated total mouse DNA.

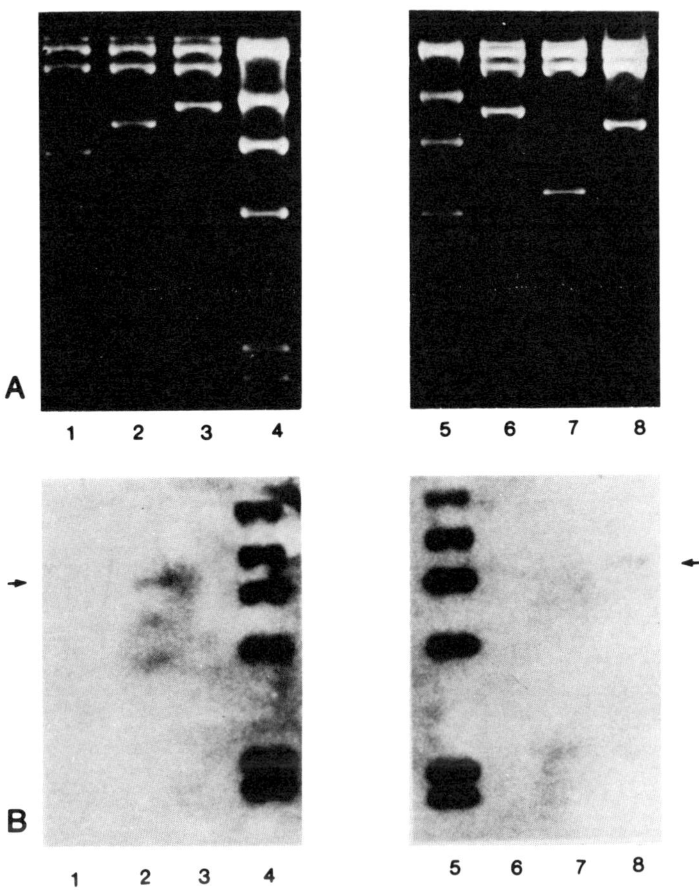

Figure 15: Analysis of DNA cloned from flow-sorted mouse chromosomes for inserts hybridizing with repetitive DNA. DNA, prepared from plaque-purified recombinant clones that were then purified in CsCl step gradients, was digested with EcoRI endonuclease. Following electrophoresis in 0.8% agarose gels, the DNA was stained with ethidium bromide for fluorescence analysis (A) and then blotted onto nitrocellulose (B). Clones were chosen after excluding those containing large amounts of repetitive DNA by Benton-Davis plaque hybridization (2) and preliminary Southern blot analysis. Lanes 1,2,3,5,7 & 8 contain DNA from phage clones #33,34,40,74,75 & 82 respectively, while lanes 4,5 and ^{32}P-labeled standards consisting of phage λ with Hind III endonuclease.

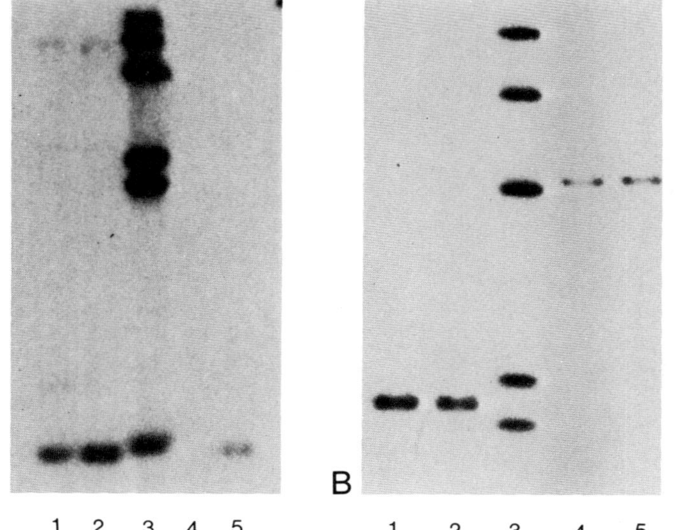

Figure 16: Southern blot analysis of cloned DNA probes for mouse X-specificity. DNA from C3H mouse liver (male, lane 1; female, lane 2). From CHO cells (lane 4) and mouse-CHO cell hybrid retaining all or most of the X as the only mouse chromosome (lane 5), were digested with EcoRI restriction endonuclease, electrophoresed in 0.8% agarose gels, blotted, and probed with different ^{32}P-labeled cloned DNA inserts. ^{32}P-labeled Hind III endonuclease-digested phage λ DNA is included in lane 3 as a size standard. In (A), the probe used, a 0.5 kb fragment derived from clone 75, exhibited mouse X-specificity, as evidenced by more intense hybridization with female DNA than with male DNA and by hybridization with DNA from the CHO X mouse cells but not with DNA from the CHO cell parent. In (B), a 2 kb fragment from clone 51 showed neither X-dependent dosage or selectivity for the CHO-mouse hybrid cell. It did, however, hybridize with a 5 kb CHO DNA fragment (Disteche et al, Cytometry, 1982 (42)).

Figure 17 shows the hybridization pattern of a slightly repeated insert that hybridizes more with female than with male DNA, and hybridizes with DNA from the hybrid containing the mouse X but not with the DNA from the hamster cells. _In situ_ hybridization experiments are planned to examine more closely the degree of X-specificity of this probe.

Thus far, out of approximately 70 plaques from the phage library enriched for X(7) sequences that did not hybridize appreciably with nick translated total mouse DNA, 32 exhibited inserts that were repeat negative on Southern blots. Twenty-three of these have been screened. Some still had repeats, while two did not hybridize with our DNA panel at all. Of 15 that were scored, 9 were not from the X. Five including two moderately repeated sequences, exhibited X-specificity, and one, which we haven't characterized further, hybridized with female mouse DNA but not with male DNA. Strain differences between DNA samples may underly this last result. Dr. Disteche, now at the University of Washington, Seattle, is beginning to isolate and characterize additional probes from this mouse library.

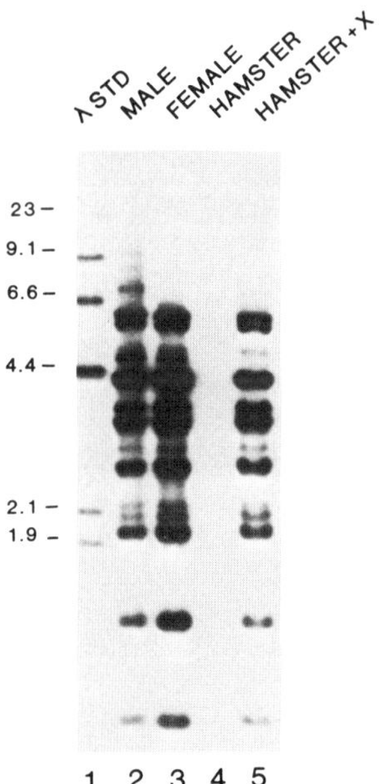

Figure 17: Southern blot of cloned mouse DNA exhibiting moderate sequence repetition and · some X-specificity. The phage insert, labeled with ^{32}P, was used to probe a blot in which gel lanes contain EcoRI digested DNA from the sources shown.

Our laboratory has now focused more on the human X, utilizing DNA from a number of rodent-human hybrid cell lines provided by Dr. Bruns and Dr. Gerald. Some of these hybrids had previously been characterized cytologically by BrdU-dye techniques. Figure 18 shows chromosomes from cells with an X-13 translocation in which the break point on the X is near band Xq22. The normal X was late replicating in all but 1-2% of these cells (47), which were derived from a 12 year old girl whom Dr. Gerald observed to exhibit some features of muscular dystrophy. Unfortunately, it was not possible to obtain additional data on this patient. Note in particular that the break point on the X is on the long arm. Cells from this patient were used by Dr. Bruns to form a hybrid line which eventually retained the translocation product containing the bottom part of the X; DNA from this hybrid was used in DNA mapping. Figure 19 shows chromosomes from cells with an X-19 translocation with a break point near Xq24 (47). DNA from this hybrid line has also been employed in subsequent studies. Figure 20 shows late replication analysis of a lymphocyte from another patient, referred

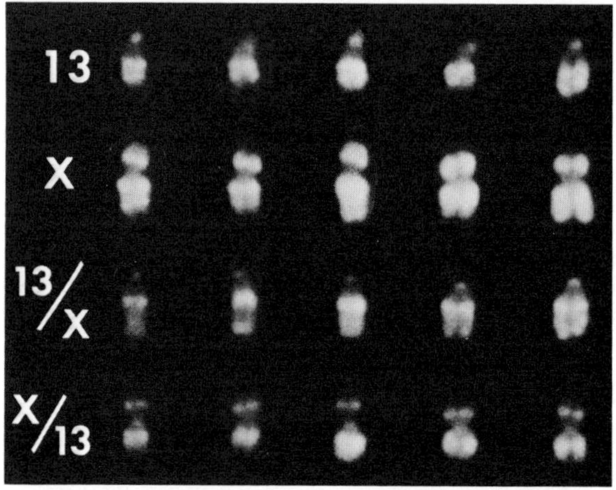

Figure 18: Late replication in #13, X, and the X-13 translocation products from a 46,X rcp(X;13) patient. Lymphocytes were grown so that BrdU was incorporated in the early S phase and stained with 33258 Hoechst. The chromosomes in each column are from the same cell. Bright fluorescence is evidence of late replication. For all cells shown, the structurally normal X was late replicating, as judged from its fluorescence (Latt et al, 1976) (47).

by Dr. Robert Greenstein, with an X-11 translocation (48). The
break point in this translocation is the short arm of the X. Fibroblasts from this patient, who has Duchenne muscular dystrophy, are
stored at the Camden Cell Repository (GM1695).

Figure 21 shows another type of cell characterized by Dr. Uma
Tantravahi and used in analysis for X-specific clones, in which
there is an isochromosome X. Lymphoblasts with this karyotype
have three copies of the long arm of the X but one of the short arm.
A somewhat converse situation, shown in Figure 22, is an X-X translocation in which there is a normal X plus a dicentric chromosome
with two copies of the short arm and all of the long arm up to band
Xq24. DNA from these cells have served to obtain results with hybrid
cells. Finally, a cell from the line that Lou Kunkel has used for
flow sorting, shown in Figure 23, is one in which there are four X
chromosomes, one early and three late replicating X chromosomes.
Here the number of copies, rather than a difference in size of the
X chromosomes is exploited, both as a source of X chromosomes and
as a source of DNA to screen probes by hybridization dosage.

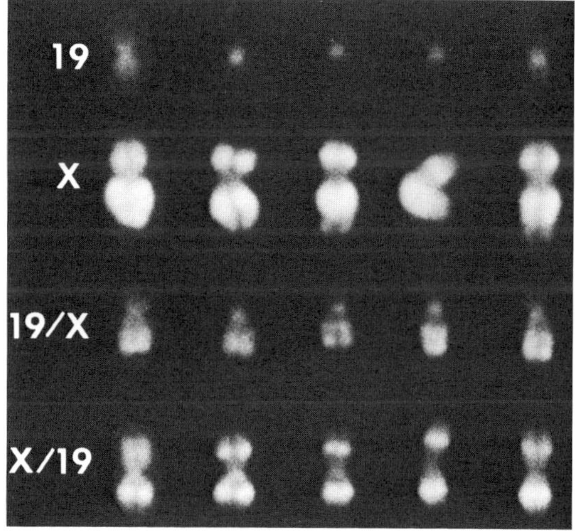

Figure 19: Late replication in #19, X, and X-19
translocation products from a 46,X rcp(X;13)patient.
Lymphocytes were grown so that BrdU was incorporated
in early S and stained with 33258 Hoechst. The chromosomes in each column are from the same cell. Bright
fluorescence is evidence of late replication. For all
cells shown, the structurally normal X was late replicating, as judged from its fluorescence (Latt et al,
1976) (47).

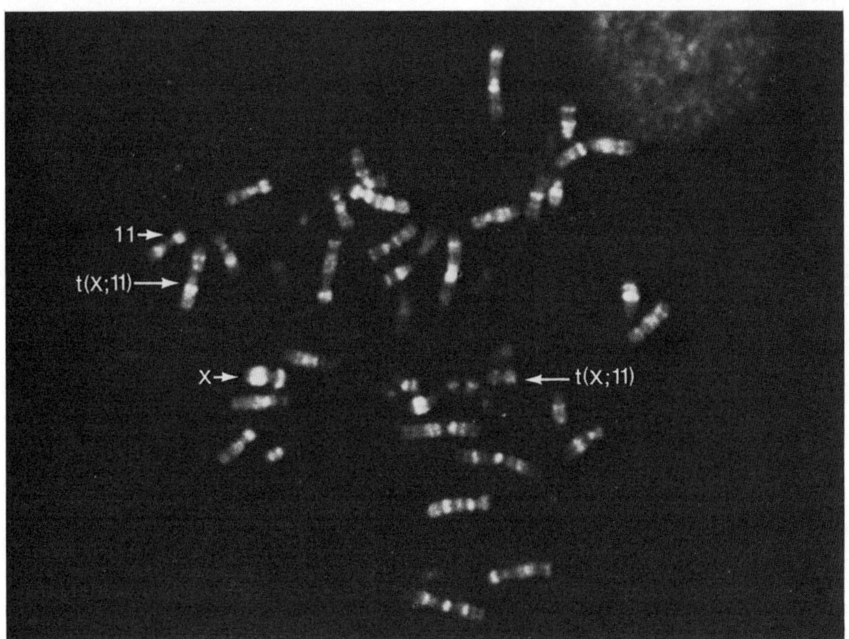

Figure 20: Late replicating regions in a lymphocyte from a patient with a rcp(X;11) karyotype.
Cells were cultured and stained as described in
the previous two figure legends. Arrows indicate
the normal X and 11 and the translocation products.

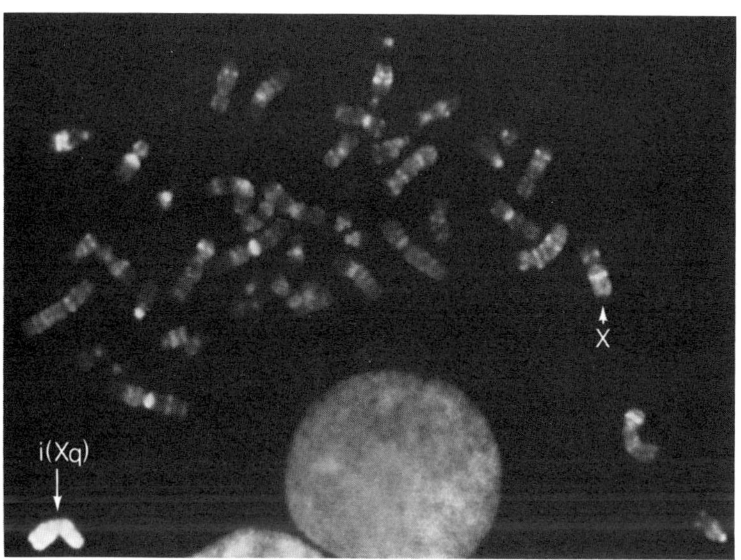

Figure 21: Late replication in a human lymphoblast containing an isochromosome for the long
arm of the X. This cell was cultured in medium
containing 10^{-4} M BrdU for 12 hours, which was
then changed to medium containing 10^{-5} M dT,
but not BrdU, 7-1/2 hours prior to harvest.
Chromosomes were stained with 33258 Hoechst;
late replicating regions exhibit relatively
bright fluorescence. The i(Xq) is indicated
by a long arrow, the normal X by a short arrow.

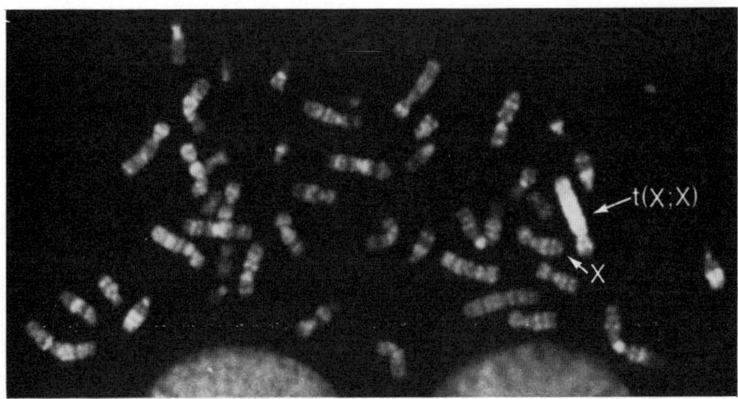

Figure 22: Late replicating regions of a human lymphoblast cell containing an X-X translocation (t(Xq24;Xq24). Cell culturing and staining was as described in the legend to the previous figure. The dic X(q24;q24) is indicated by a long arrow and the normal X by the short arrow.

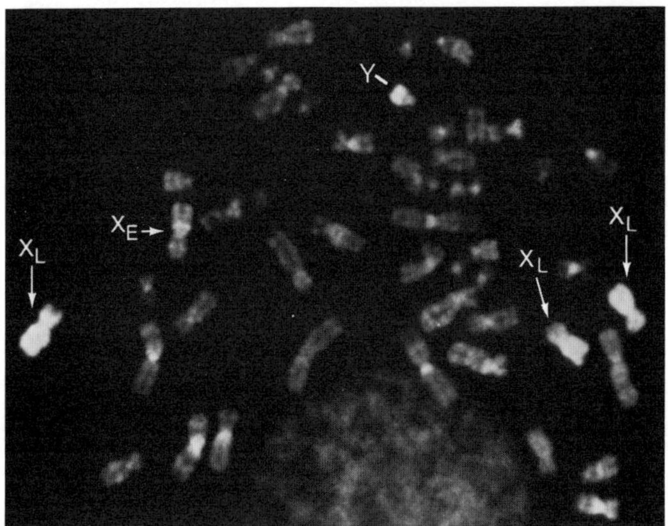

Figure 23: Late replicating regions in a 49, XXXXY lymphoblast. The late replicating X chromosomes are indicated by long arrows and the early replicating X chromosome by a short arrow.

DR. WHITE: The X-specific repeat has been tested by dosage for being X-specific. Has it been tested against a set of mouse autosomes under conditions where X is not present?

DR. LATT: No. The evidence for X-specific hybridization consists of differential hybridization with male mouse DNA vs. female mouse DNA and with hamster plus mouse X DNA vs. hamster DNA. At the very least, in situ hybridization, not yet done, will be needed to exclude hybridization with autosomes.

DR. HECHT: How do you resolve the active from the inactive X.

DR. LATT: There are at least two ways. With the Cattanach X it is easy. The X(7) is much larger than the normal X. Dr. Disteche has established embryo lines in which she could show cytologically that the X(7) was either late or early replicating. Similar resolution should be possible with material from human cells containing large, structurally abnormal X chromosomes.

DR. FRANCKE: Do I interpret your answer correctly that you can distinguish structurally abnormal Xs due to the fact they are larger, but you can't distinguish two structurally normal Xs on the basis of activity being active or inactive?

DR. LATT: No. Dr. Marc Lalande in our laboratory has initiated experiments directed at flow sorting chromosomes from synchronized lymphoblast lines. If such cells are pulsed with BrdU before harvest, the early and late replicating X chromosome should be resolvable because of their differential 33258 Hoechst fluorescence (49). This approach might further be refined by using a second stain e.g. ethidium bromide, which does not exhibit BrdU-sensitive fluorescence, and sorting chromosomes according to the ratio of 33258 Hoechst to ethidium fluorescence (50).

DR. SINISCALCO: I wonder why hybrids with the individual chromosome haven't been used as yet as a source for isolating specific chromosomes with the flow.

DR. LATT: It has been discussed and should be feasible.

DR. SINISCALCO: Because the contamination can be easily pulled apart.

DR. LATT: Correct.

DR. WILLIAMSON: The limiting thing here is collecting the
amounts of material that you need to clone. It takes us 12 hours
to get a microgram of DNA with a XXXX human cell line. We need
at least half to 3/4 of a microgram to clone. If you use a rodent-
human hybrid cell that has got one X in it, that means you are
running the machine for a week. We don't usually get our machines
to work for a week at a time. I don't know why we don't do it that
way because you can do it with a XXXX cell line in a day.

DR. LATT: Another point is that it might not be easy to
prepare metaphase chromosomes from all hybrid lines in high yield.

DR. SINISCALCO: Maybe it is still worthwhile to think about
it because, apart from everything else, the series of X autosomal
translocation with one piece of the X only and one piece of a
specific autosome might open the way to construct libraries for
those particular regions which might be useful to have available.

DR. LATT: In principle, this is true. Its practical utility
has not yet been established. Thus far, the strategy has been to
enrich for X-specific sequences and then to assign them to sub-
chromosomal regions by Southern blots. Alternatively, one might
start with sequences which, if from the X, are restricted to a
region present on a translocation. The methods are complementary
and might be subject to (or free of) different sets of complications.

CHAPTER 3

DNA PROBES: DEVELOPMENT OF LIBRARIES

DR. LOUIS M. KUNKEL: Dr. Latt has described some of the work
currently in progress in our laboratory but has left to me the task
of describing our work related to flow sorting and cloning the human
X chromosome. The goal was to obtain a library of recombinant phage
highly enriched for X chromosome DNA sequences. These phage would
serve as sources of DNA possibly useful in linkage analysis with
known X-linked loci or in the acquisition of the more than 100
known X-linked genes (52). I would like to describe briefly the
construction of the X chromosome enriched library and our initial
characterization of some of the human inserts from this library.
Lastly, I would like to describe our initial efforts to obtain
sequences of DNA which might possibly encode for X-linked proteins.

Fluorescence-activated chromosome sorting has been employed
by our own laboratory (42,43) as well as by others (53-55) to obtain
a fraction enriched for a particular chromosome. We chose a 49,
XXXXY lymphoblast cell line (GM 1202, from the Camden Cell Repository, Camden, N.J.) as our source of human chromosomes for preparative sorting. Shown in Figure 24 are two different flow histograms
of chromosomes stained with 33258 Hoechst. Shown in the upper
panel is the flow histogram of chromosomes derived from a 46,XY
lymphoblast cell line. The chromosome designations are taken from
similar published histograms (54). The lower panel shows the same
flow histogram of chromosomes isolated from a 49,XXXXY cell line.
As indicated by the arrow, the X peak has increased at least two-
fold in height and should reflect the increased number of X chromosomes found in this 49,XXXXY cell line. Over the course of a few
weeks, chromosomes from the X containing peak were preparatively
sorted and stored as frozen.

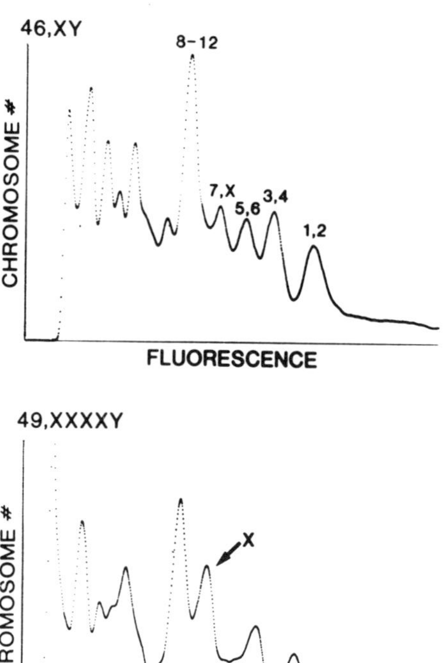

Figure 24: DNA flow histograms of human lymphoblast chromosomes stained with 33258 Hoechst.
Upper panel is a reproduction of a histogram of
46,XY chromosomes displayed by the ND-100 multi-
channel analyzer of the Becton-Dickinson FACS II
sorter. The chromosome designations are as described. The lower panel is a similar reproduction
of 49,XXXXY lymphoblast chromosomes. The arrow
denotes the increased peak containing X chromosomes
which was preparatively sorted.

Shown in Figure 25 is a summary of the chromosome sorting and cloning. From approximately 10^7 sorted chromosomes, $0.6\,\mu g$ of DNA was isolated. Cloning of a portion of this DNA was accomplished in the Hind III λ phage vector Charon 21A (56). Approximately 80-90 percent of a total Hind III restriction endonuclease digest should fall within the packaging range of this phage. One hundred nanograms of Hind III cleaved chromosomal DNA was ligated with 50 nanograms of Charon 21A DNA that had been cleaved with Hind III endonuclease and treated with calf intestinal alkaline phosphatase. The resulting 60,000 in vitro packaged phage were amplified in lawns of LE392 bacteria and the phage stored. This Hind III library should complement with the EcoRI library of human X chromosome DNA to be presented soon by Dr. Williamson (55).

Initial experiments with the X chromosome library were general in nature. An aliquot of phage was plated on lawns of LE392 bacteria and the resultant plaques screened by Benton-Davis analysis (51) for the presence of human repeated DNA sequences.

Chromosomes sorted. . . . . 10×10^6

X content 30-60%

DNA isolated. . . . . 0.6 ug

Vector. . . . . λ Charon 21A

Library size. . . . . >60,000 pfu

Average insert size . . . . 3-4 kb

Inserts lacking repeated sequences. . . <20%

Screening . . . . X specificity

                              . . . . . Subchromosomal location

                              . . . . . Expression

                              . . . . . Linkage

Figure 25:   A summary of the cloning of
human X-specific DNA from flow sorted chromosomes.

More than 80 percent of the plaques were homologous to total human
DNA labeled by nick translation with ^{32}P (57). The plaques which
were free of human repeated DNA sequences were picked and amplified
and the DNA was isolated. Analysis of purified phage DNA by cleavage
with Hind III yielded inserts of DNA ranging in size from 0.5 to 9 kb
with a mean size of 3 to 4 kb.

Once having established the human nature of the inserts from
the library, those which were free of repeated sequences were tested
for X chromosome specificity by both
Human X-Specific Inserts qualitative and quantitative Southern
hybridization analyses (43,58). The
qualitative analysis relied on differential hybridization to rodent-
human hybrid cell DNA samples with either an intact human X chromosome or varying fragments of the X within a background of rodent
chromosomes. The quantitative analysis utilized differences in
hybridization intensity to various human DNA samples with structurally
abnormal or supernumerary X chromosomes. Shown in Figure 26 are such
analyses for two inserts from the X chromosome library. The first is
thought to derive from X chromosome short arm and the second from the
distal end of X chromosome long arm. The DNA samples were cleaved
with either EcoRI or Hind III in preparative quantities and an
aliquot of each restriction endonuclease digest was subjected to
electrophoresis in agarose gels. Following the separation by
molecular weight, the fragments were then denatured, transferred to
a nitrocellulose membrane, and the immobilized DNA probed by DNA
hybridization with a ^{32}P-radiolabeled insert from the library (58).
The right and left panels in Figure 26 are autoradiographs of the
resultant hybridized nitrocellulose filters. The panel on the right
is the hybridization of an insert thought to derive from the short
arm of the X chromosome. This 1.2 kb Hind III insert hybridizes
with a 3.2 kb EcoRI fragment in the rodent-human hybrid DNA with an
intact human X chromosome (lane 4). No hybridization is detected in
lanes 3 and 5 which contain hybrid DNA samples containing human X
chromosome fragments of long arm representation but no short arm
material. The short arm localization is substantiated by the Hind III
digests of various human DNA samples. A 1.2 kb fragment hybridizes
more strongly to DNA with two chromosomes (lane 9, Figure 26) than to
DNA with one X chromosome (lane 10) and much more strongly to DNA with
four X chromosomes (lane 6). Hybridization to DNA derived from cells
with karyotype 46,X dic(Xq) (lane 7) was intermediate in intensity
between 46,XX DNA (lane 9) and 49,XXXXY DNA (lane 6). The dicentric
chromosome is such that there are three copies of X chromosome short
arm along with two-thirds of the long arm. Critical to the X chromosome
short arm localization is the similarity of hybridization intensity
observed between the 46,XY DNA (lane 10) and the 46,X,i(Xq) (lane 8).
The isochromosome-containing cell line has three copies of the long
arm of the X chromosome and one copy of X chromosome short arm. From
both EcoRI cleaved hybrid DNA samples (lanes 3,4 and 5) and the human
cell lines cleaved with Hind III (lanes 6,7,8,9 and 10) we conclude

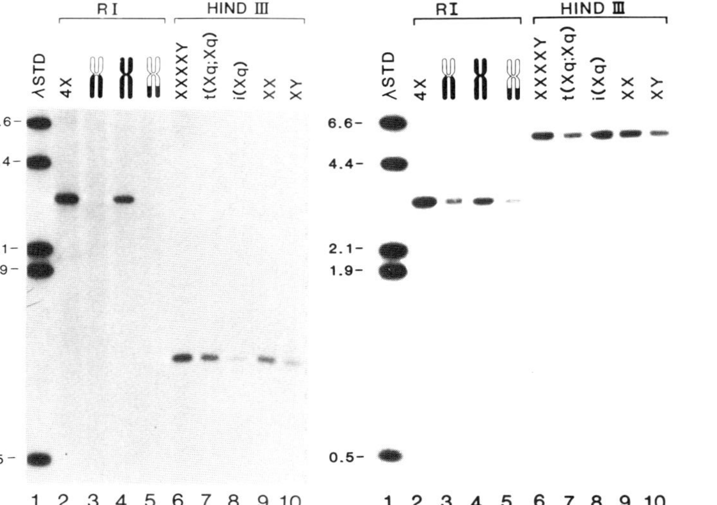

Figure 26: Localization of a 1.2 kb Hind III insert to the X chromosome short arm and a 5.5 kb Hind III insert to the X long arm. The right and left panels are autoradiographs of the hybridization pattern of a ^{32}P-labeled insert to equal aliquots of restriction endonuclease digests of DNA immobilized on nitrocellulose (58). The inserts were radiolabeled by nick translation (59) as subcloned inserts in pBR322 (57). Three μg of DNA from each sample was loaded in the following order: lane 1, ^{32}P-labeled Hind III cleaved λ DNA; lane 2, EcoRI cleaved 49,XXXXY lymphoblast DNA; lane 3, EcoRI cleaved mouse-human hybrid cell DNA with most of the human X chromosome long arm; lane 4, EcoRI cleaved hamster-human hybrid cell DNA with an intact human X chromosome; lane 5, EcoRI cleaved mouse-human hybrid cell DNA with the distal half of the X chromosome long arm (Xq22→Xqter); lane 6, Hind III cleaved, 49,XXXXY lymphoblast DNA; lane 7, 46,X,dic(X) (q24) lymphoblast DNA with the translocation chromosome being deleted for Xq24→Xqter; lane 8, Hind III cleaved 46,X,i(Xq) lymphoblast DNA; lane 9, Hind III cleaved 46,XX lymphoblast DNA; lane 10, Hind III cleaved 46,XY lymphoblast DNA. Above each lane is a schematic representation of the X chromosomes present in each cell line. The 1.2 kb Hind III insert hybridized in the right panel is thought to derive from X chromosome short arm. The 5.5 kb Hind III insert hybridized in the left panel is thought to derive from X chromosome long arm Xq24→Xqter.

that this 1.2 kb insert from the library is localized within the region Xpter ->Xcen.

Shown in the left hand panel of Figure 26 is the hybridization of a 5.5 kb Hind III insert to aliquots of the same samples as described in the right panel. This 5.5 kb Hind III insert hybridizes to a 3.5 kb EcoRI fragment in all three rodent-human hybrid cell DNA samples (lanes 3,4 and 5). The only portion of the human X chromosome common to all three samples is that between Xq22 and Xqter. Hence, this fragment is tentatively localized to the distal end of X chromosome long arm. The long arm localization is substantiated by the results presented in the left panel with Hind III digests of human cell line DNA samples (lanes 6 through 10). The pertinent results are the hybridization intensities observed in lanes 7 and 8 for a 5.5 kb fragment. The 46,X,i(Xq) DNA separated in lane 8 has three copies of the long arm of the X chromosome and yields an intensity of the 5.5 kb fragment between that for 46,XX DNA (lane 9) and 49,XXXXY DNA (lane 6). The dicentric chromosome present in the 46,X, dic(Xq) (lane 7) is such that there are three copies of X chromosome Xpter->Xq24 and only one copy from Xq24->Xqter. The hybridization intensity observed for the 5.5 kb Hind III fragment on this DNA sample is similar to that observed for 46,XY DNA (lane 10). From results with both EcoRI digests of hybrid DNA samples (lanes 3,4 and 5) and Hind III digests of various human cell lines (lanes 6,7,8, 9 and 10) we feel that this 5.5 kb insert from the library is localized to X chromosome long arm Xq24→Xqter.

Thus far, eleven inserts from the library which are free of repeated sequences have been tested for X chromosome specificity and of

Isolation of X Chromosome DNA Sequences

these, seven have been localized to the X chromosome. Hence, the library represents a source of DNA sequences of which approximately 65 percent are derived from the X chromosome. Data subsequently obtained on 22 additional inserts substantiate this estimate. Shown in Figure 27 is a summary of the chromosomal localizations for these seven library inserts. Each sequence has been localized by the dual analyses depicted in Figure 26. The hybridization analyses described here allow localization to four regions of the X chromosome. Further subregional localization can be accomplished by utilizing the various cell lines described at this meeting by Drs. Francke, Bruns, Siniscalco and Williamson. The seven inserts described here represent the beginning of a structural DNA sequence map of the human X chromosome. If restriction fragment length polymorphisms (36) can be found for these DNA sequences, then linkage analysis can be attempted within families segregating an X-linked disorder thought to arise from a particular section of the X chromosome (such as Duchenne's muscular dystrophy (DMD) or hemophilia). We are currently attempting to find such polymorphisms surrounding our regionally localized DNA fragments.

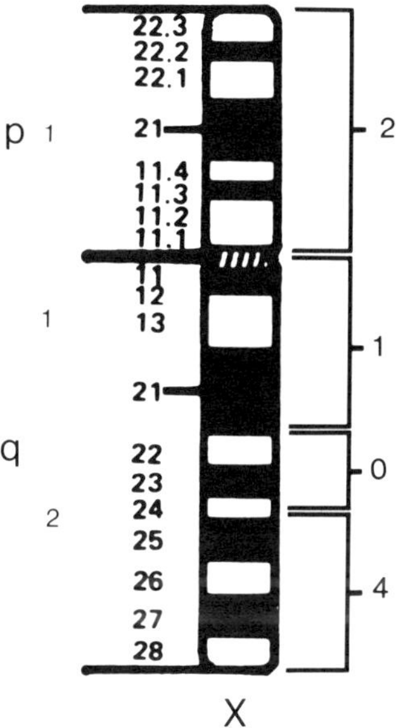

Figure 27: Schematic representation of localization for seven X chromosome specific sequences.
Seven of eleven inserts from the library are X
chromosome specific and have been localized to
various sections of the X chromosome by analyses
described in Figure 26. They are schematically
mapped in this figure.

We then hope that these will be segregating within families known
to be at risk for DMD or other X linked disorders.

I would like to close with a description of some preliminary
experiments we have started in collaboration with Dr. David Kurnit.
The objective is to isolate DNA sequences from the X chromosome
which encode for RNA which might ultimately be translated into
protein. Towards this goal approximately 1000 cDNA clones derived
from HeLa cell mRNA were prepared separately as DNA. One hundred
of the clones which were free of repeated DNA sequence transcripts
were pooled together and radiolabeled with ^{32}P utilizing T4 DNA
polymerase (60). This pool of plasmids was then used as a probe
against replica filters of immobilized phage from the human X
enriched library. Shown in Figure 28 is an autoradiograph of duplicate filters in which four plaques on both filters at the same
positions are homologous to the mixed cDNA probe. The positive
plaques were recovered separately, replated and rescreened with
the same pool of DNA plasmids. Shown in Figure 29 is an autoradiograph of duplicate filters from one rescreening. Large numbers
of presumably the same original phage are now homologous to the
mixed cDNA probe. DNA has been prepared from some of the phage
thought to be homologous to these HeLa cell cDNA clones. The X
chromosome specificity of the phage DNA inserts is currently being
tested. We assume that due to the enriched nature for the X
chromosome of the library that some of these phage inserts homologous to cDNA are derived from the human X chromosome.

The X chromosome enriched library presented here as well as
the one presented by Dr. Williamson should yield numerous fragments
of DNA from the human X chromosome. Whether polymorphic or expressed, these should serve as markers for a physical and functional
map of the entire X chromosome.

    * DR. GAIL BRUNS: Among the unique features of the human X
chromosome are its pairing characteristics at meiosis, the dosage
compensation mechanism of inacti-
Repetitive Deficient   vation of many loci in somatic cells
X-Specific Probes   of the female, the conservation of
the genetic content of the chromo-
some in diverse species, and the assignment of at least 100 loci
to the chromosome, including, presumably, the locus for Duchenne
muscular dystrophy.

    * Work done in collaboration with J.F. Gusella of the
       Massachusetts General Hospital, D. Housman, Massachusetts
       Institute of Technology and P.S. Gerald of Children's
       Hospital Medical Center, Boston

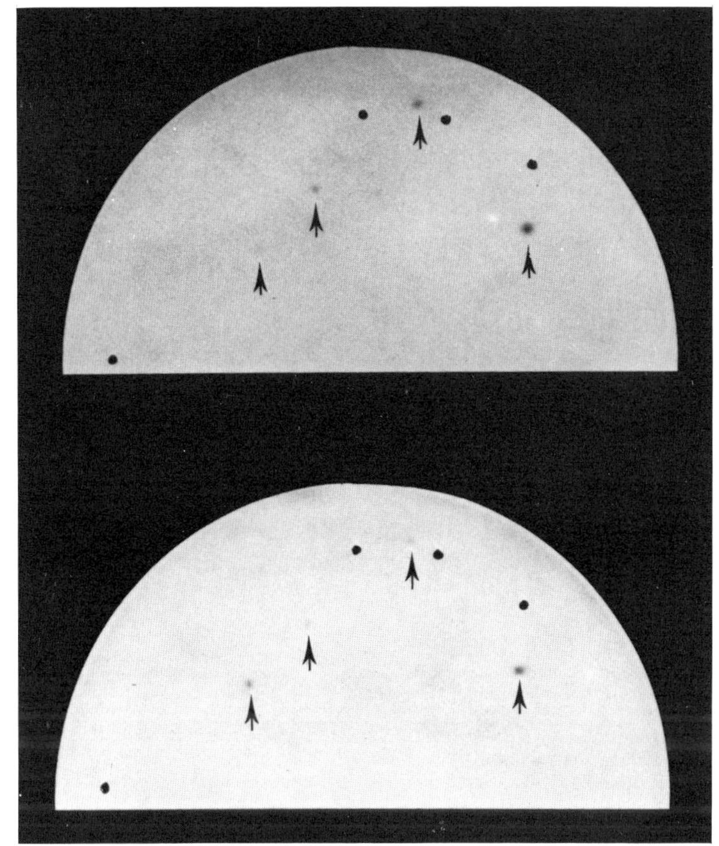

Figure 28:   Primary Benton-Davis analysis of pooled
HeLa cell cDNA probes. Approximately 50,000 pfu of
the enriched X-library were plated on 5 plates with
lawns of LE392 bacteria. The resultant plaques were
transfered to duplicate nitrocellulose filter discs
by the procedure of Benton and Davis (51). A pool of
100 cDNA clones derived from the HeLa cell mRNA
was radiolabeled by utilizing T4 DNA polymerase
(60) and hybridized to the replica filters. Each
cDNA clone was previously shown not to hybridize to
^{32}P-labeled human repeated sequences. The four
arrows in both duplicates, A and B, point to positive
hybridization to four single plaques in identical
positions on the duplicate filters. The autoradiograph presented here was exposed for 3 days.

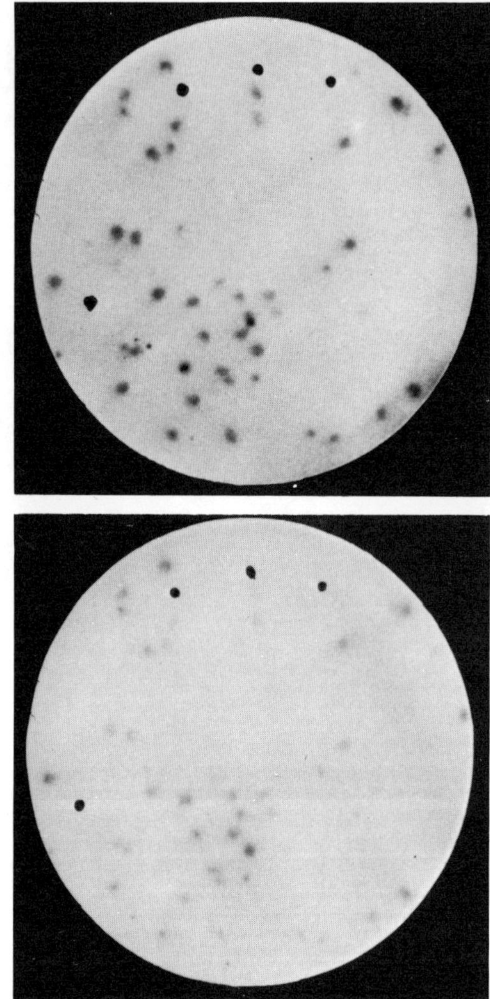

Figure 29: Secondary plaque screening with pooled
DNA probes. One of the positive plaques from the
primary screen shown in Figure 28 was picked and
reamplified on a lawn of LE392 bacteria. The resultant phage plaques were diffused onto duplicate
nitrocellulose filters and probed once again with
the pool of HeLa cell ^{32}P-cDNA clones. Approximately
20 percent of the plaques now hybridize with the ^{32}P-
probe.

For current and future studies of this chromosome, we have isolated a number of repetitive deficient DNA segments from the human X by methods developed by Gusella et al. (33). These segments have been obtained from a recombinant library prepared in the bacteriophage Charon 4A with DNA from a human-hamster somatic cell hybrid that contained the X and chromosome 20 as the only human chromosomes. Ten DNA segments derived from the X have been localized to one of four regions of the chromosome. Eight of these segments appear to recognize sequences that may be specific to this chromosome. Here we summarize the procedures used in the isolation of these recombinants and record the initial characterization of some of the DNA segments. Part of this work has been previously communicated in abstract form (40).

For preparation of the recombinant library, DNA was isolated from a human-hamster somatic cell hybrid that had, as the only human complement, chromsomes X and 20. No other human chromosomes or human-hamster translocations were detected by sequential quinacrine mustard and Giemsa 11 staining of the metaphases. The hybrid expressed the human isozymes characteristic of these two chromosomes and did not express 36 additional isozymes assigned to the other 21 autosomes (61).

A partial EcoRI digest of the DNA from the hybrid was fractionated according to size by sucrose gradient centrifugation to enrich for 12-20 kb segments (62). The latter were ligated to purified Charon 4A arms and packaged _in vitro_ into viable bacteriophage (56). The recombinants were propagated on the bacterial host LE392 at P-1 EK-1 containment.

To identify phage with human DNA inserts, isolated plaques were transferred to gridded bacterial plates (51). Two nitrocellulose filters were prepared from each grid: one filter was hybridized with radiolabeled HeLa DNA and the second, with comparably labeled hamster DNA under conditions optimized for hybridization of moderately repetitive sequences including the Alu family of DNA sequence (65). By this means, plaques that hybridized only with HeLa DNA were identified. Such plaques, which presumably represented phage with human DNA inserts, were isolated and plaque purified prior to DNA preparation.

The DNAs were digested with one of four restriction endonucleases (EcoRI, Bam HI, Hind III and Sac I) and the resultant DNA segments visualized on agarose gels by ethidium bromide staining. The digested DNAs were transferred to nitrocellulose filters by the technique of Southern (58) and hybridized with HeLa DNA labeled by nick translation (S.A. 2 x 10^8 cpm/μg) (57). DNA segments of insert origin that did not hybridize with HeLa DNA were identified and subsequently isolated from preparative agarose gels by electroelution for use as probes (63). Such DNAs were designated "repetitive deficient" DNA segments.

The chromosomal and regional assignment of each repetitive deficient DNA segment was demonstrated by hybridizing the radio-labeled DNA to a panel of DNAs from various somatic cell hybrids with intact or rearranged X chromosomes. DNAs from human HeLa cells and from the rodent parental cells for the hybrids were also included in the panel. The hybrid and parental DNAs were digested with EcoRI, electrophoresed on one percent agarose gels and transferred to nitrocellulose in 20X SSC. Comparable amounts (5-7 µg) of hybrid and parental DNA, determined by quantitation of DNA fluorescence with DAPI (64), were loaded on the gels. Nitrocellulose filters were hybridized with 4-10 x 10^6 cpm of labeled probe in 6X SET (0.9 M NaCl, 0.012 M NaEDTA, 0.18 M Tris HCl, pH 8) at 60°C with 10 percent dextran sulfate or at 65°C without dextran sulfate (33). Filters were washed at the same temperature as that used for hybridization with decreasing concentrations of SET to a final concentration of 0.3X SET for filters hybridized at 65°C or 0.1X SET for those hybridized at 60°C. Autoradiograms were exposed for 2-5 days at -80°C with an intensifying screen.

Fifty phage clones with human DNA inserts have been isolated from the recombinant library prepared in Charon 4A with DNA from the somatic cell hybrid with human
Cloned Human DNAs   chromosomes X and 20 as the only
human complement. All of these clones were identifed by virtue of the hybridization of repeat DNA sequences of the insert with members of the repetitive DNA families of HeLa DNA. None of the clones demonstrated significant hybridization with repeat DNA sequences of hamster DNA. To determine whether hybridization of the cloned human DNA segments was due, in part, to sequences of the Alu I repeat DNA family, a subset of 15 clones (13 from the X and 2 from chromosome 20) was hybridized by a Benton – Davis procedure (51) to a sequence (Blur 11) which is a member of this family cloned in pBR322 (65). All 15 clones hybridized with this sequence (Figure 30).

The DNAs from 28 clones (including those hybridized with the Blur 11 probe) have been analyzed for repetitive deficient DNA segments by restriction endonuclease digestions and subsequent hybridization of the digestion fragments with the repeat families of HeLa DNA (Figure 31). Such segments were identified in 21 of the DNAs with one of four restriction endonucleases (Bam HI, EcoRI, Hind III and Sac I).

For the initial assignment of the repetitive deficient DNA segment from these phage to either the X chromosome or chromosome 20, the segment was isolated from a preparative agarose gel, labeled by nick translation, and hybridized with a panel of DNAs from a number of cell types. These included: a) the somatic cell hybrid used for the preparation of the library that had only human chromosomes X and 20; b) a somatic cell hybrid with only the X and an

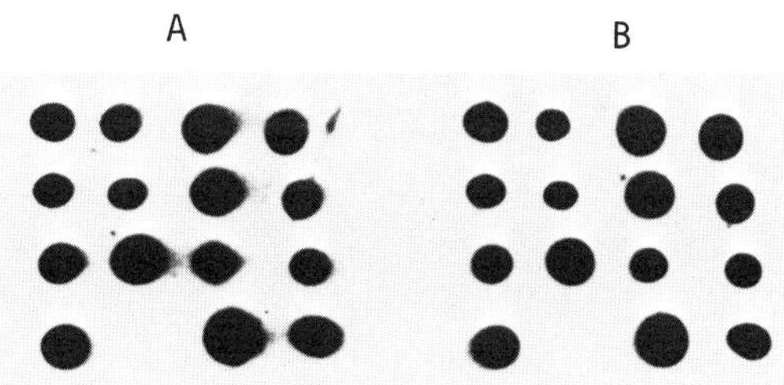

Figure 30: Primary phage stocks were spotted onto
a lawn of LE392. Plates were incubated overnight
at 37°C and duplicate Benton-Davis filters (51)
prepared. One filter (A) was hybridized with
5×10^6 cpm of nick translated HeLa DNA and the
second filter (B), to 5×10^6 cpm of nick translated Blur 11 DNA. The phage of the top three
rows and the first column of the fourth row are
derived from the X chromosome, and the remaining
two phage, from chromosome 20.

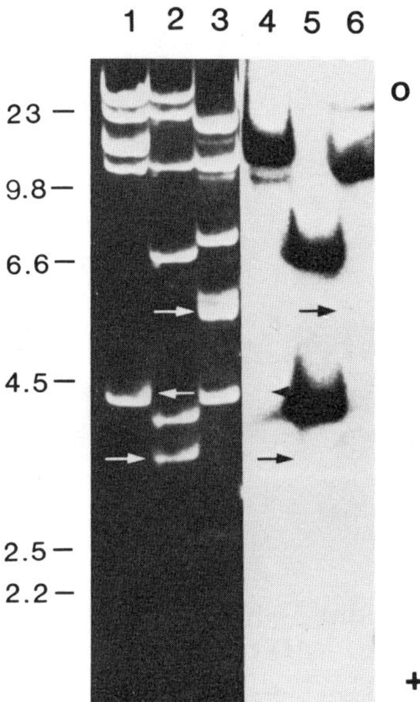

Figure 31: Comparison of the restriction digestion
patterns of three X chromosome derived recombinants
(lanes 1-3) with the corresponding autoradiograms
following hybridization of the digested DNAs with
labeled HeLa DNA (lanes 4-6). The white arrows
(lanes 1-3) and black arrows (lanes 4-6) indicate
non-phage bands that do not hybridize with HeLa DNA.
These repetitive deficient DNA segments were subsequently regionalized on the X chromosome. Hybrid-
izations were performed with 5 x 10^6 cpm of nick
translated HeLa DNA. A Hind III digest of lambda
DNA was used for the size markers (left).

unidentified human-hamster translocation in a small fraction of the
cells; c) human HeLa cells; d) the hamster parental cells for the
hybrids (YH21 and Wg3h); and e) the mouse cell RAG. It was expected
that a cloned DNA segment from the X chromosome would hybridize with
HeLa DNA and with that from both somatic cell hybrids whereas a
DNA segment from chromosome 20 would hybridize with DNA from the
hybrid with this autosome as well as with HeLa DNA. The degree of
cross-reactivity of a human DNA segment with the several types of
rodent DNA was also assessed. By this means, twelve different DNA
segments were provisionally assigned to the X chromosome and two, to
chromosome 20. It is to be noted that the endonuclease restriction
pattern of each of the twelve X chromosome-derived DNAs was unique,
as were those of the two clones from chromosome 20.

The repetitive deficient DNA segments from ten of the X chromosome-derived DNAs have been assigned to subregions of the chromosome
by hybridization to DNAs from somatic cell hybrids that had retained
particular regions of the X. These hybrids had been prepared by
fusion of HPRT deficient hamster or mouse cells (E36, YH21, Wg3h,
RAG) with white blood cells or fibroblasts from carriers of balanced
X-autosome translocations. The regions of the X retained in the
several hybrid series is shown in Figure 32. These translocation
hybrids permit division of the X into four subregions.

The initial panel used for regional assignment of an X chromosome-derived DNA segment included DNAs from: a) human HeLa cells;
b) the rodent parental cells of the hybrids; c) three somatic cell
hybrid clones with an intact X; and d) two independent hybrid
clones with each subregion of the X. Subsequent to the initial,
provisional assignment of a DNA segment to a subregion of the X,
the segment was hybridized to a second panel of DNAs for confirmation of the assignment. The confirmation panel for each region
of the X included DNAs from 2-3 hybrid clones that had retained
that segment of the chromosome as well as DNA from one or more
clones selected in azaguanine (AZ) or thioguanine (TG) for loss of
the particular region of the X. The hybrid clones used to represent
the several regions of the X in the confirmation panel were independent primary clones, not those used in the initial assignment
panel. The confirmation panel for the repetitive deficient DNA
segment of phage 19 from the Xq1 → q22 region is shown in Figure
33.

The regional localization of ten of the repetitive deficient
DNA segments from the human X is schematically represented in Figure
34. The assignment of one of the DNAs to the pter → q1 region is
provisional as is that of one of the DNAs of the q24 → qter region.
The localizations of the other eight DNAs have been observed on both
the initial and confirmation panels.

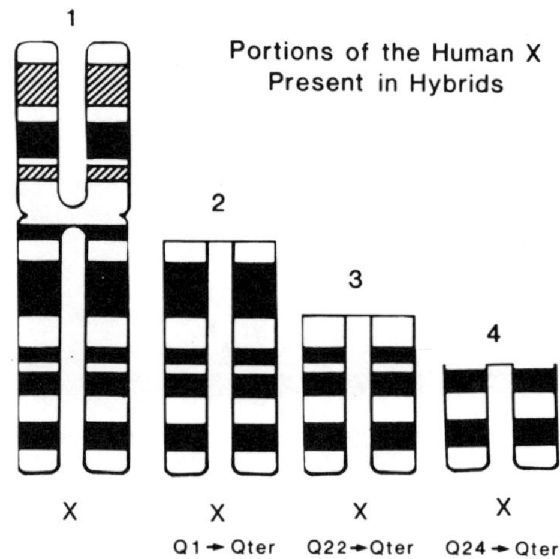

Figure 32: Schematic representation of the regions of the X chromosome present in the hybrid DNAs used for localization of the isolated X chromosome DNA segments. 1) Intact X. 2) Xq1 → qter region present in hybrids of RAG with WBC from the X-19B translocation carrier. 3) Xq22 → qter region in hybrids of RAG with fibroblasts from the X-13 translocation carrier. 4) Xq24 → qter region in hybrids of RAG or E36 with WBC from the X-19W translocation carrier. The breakpoints on the X chromosome in the X-13 and X-19W translocation carriers were determined by Latt et al (47). The cytogenetic and isozyme characterization of the somatic cell hybrid series have been previously described (61).

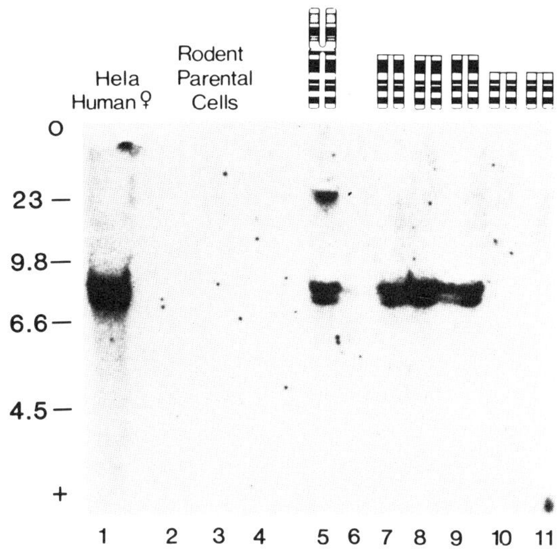

Figure 33: Hybridization of the repetitive
deficient segment from phage 19 with EcoRI
digests of human, rodent and hybrid DNAs.
Lanes: 1) HeLa. 2) E36. 3) Wg3h. 4) RAG.
5) Hybrid with an intact X and chromosome
20. 6) AZ resistant derivative of the hybrid
in lane 5; this subclone had a trace of human
PGK and the DNA hybridizes slightly with the
probe. 7-9) Hybrids with Xq1 → qter; these
lines lack chromosome 20. 10-11) Hybrids with
Xq22 → qter. The probe from phage 19 did not
hybridize with DNA from a TG resistant derivative of the hybrid in lane 7 (not shown).
This subclone lacked the Xq linked isozymes and
the Xq1 → qter chromosome. 5-7 ug of DNA from
each of the cell lines was loaded onto the gel.
The probe was labeled with ^{32}P dCTP by nick
translation (S.A. 2 x 10^8 cpm/ug). Hybridization was performed at 60°C in 6X SET with
10% dextran sulfate. The filter was washed
at 60°C in 0.1X SET. Two day exposure at
-80°C. The schematics above the hybrid lanes
show the portion of the X chromosome in the
cells. The higher molecular weight band in
lane 5 represents an incomplete digestion of
the DNA. This band was not observed in five
other digestions of the DNA from this hybrid
or in any of the other hybrid cell lines.

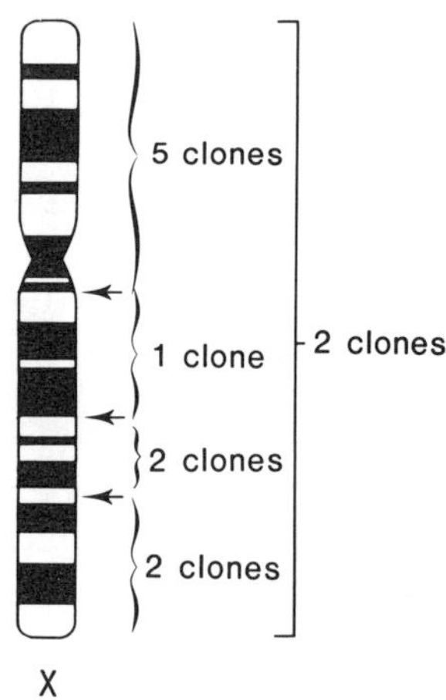

Figure 34: Schematic representation of the human X
chromosome and the localizations of the several cloned
DNA segments. The number of repetitive deficient DNA
segments localized to each of the subregions of the
X are indicated (right). The assignment of one of
the segments to the pter → q1 region is provisional
as is that of one of the segments to the q24 → qter
region. The other eight regionalizations have been
observed on both the initial and confirmation DNA
panels. Two of the cloned DNAs have not yet been
regionally localized (far right).

The human-rodent somatic cell hybrids used for preparation of
the DNA panels had been extensively characterized for their human
chromosome complements both by cytogenetic and isozyme techniques
(61). Prior to their use, an aliquot of the cell suspension used
for the preparation of DNA was analyzed for expression of the X-
linked isozymes, the isozyme(s) characteristic of the autosome
involved in the translocation, the chromosome 20 marker adenosine
deaminase, and by cytogenetic techniques for the relevant X-autosome
translocation chromosomes, the X chromosome, chromosome 20, and the
other human chromosomes in the metaphases.

The consistent segregation of the hybridization signal of a
repetitive deficient DNA segment from the X,20 library with a particular X chromosome region is presumptive evidence for the derivation
of the segment from that region of the X. If additional copies of
the sequence were present on other chromosomes which were segregating
independently of the X in the hybrids, it would be expected that the
hybridization pattern would not show concordant segregation with
the subregion of the X chromosome. For the X chromosome derived
DNA segments, examination of the human chromosome complement of hybrids whose DNA did not hybridize with a particular repetitive deficient fragment, and of the chromosomes absent in those that did
hybridize provides a segregation analysis of the fragment versus
the autosomes. In addition, analysis of the chromosomes present in
hybrids that did not hybridize with the DNA segment permits the
development of an exclusion map of the fragment from the autosomes.
Both types of analysis have been carried out with eight of the twelve
X chromosome derived DNAs. These DNA segments appear not to hybridize
with major sequences present on the autosomes. The specificity of
the X chromosome assignments of these DNAs was also demonstrated with
DNAs from the hybrid cell lines selected in AZ or TG for the loss of
each region of the X. Two of the remaining four clones hybridize as
single bands with DNAs from X-containing hybrids but have not yet
been completely analyzed for recognition of autosomal sequences.
The third clone may be a member of a multigene family with one or
more representatives on the X and the fourth hybridizes with a
component on the X and with a component of different molecular
weight on an autosome.

To provide additional evidence for the X chromosomal derivation
of the cloned DNA segments, gene dosage experiments were carried out
with seven of the segments regionally localized on the X. These
segments were hybridized to a panel of Hind III or Msp I digested
DNAs of lymphoblast origin from three individuals with different
numbers of X chromosomes; an XY male, a 48 XXXX female, and a 49
XXXXY male. Comparable amounts of the digested DNAs from each
were loaded on the gels. All seven cloned DNA segments demonstrated several fold more intense hybridization with the multiple X
DNAs than with that of the XY male. Both of the DNA segments from

chromosome 20 hybridized with equal intensity with all three DNAs,
as expected.

The two DNA segments that had been tentatively assigned to
chromosome 20 were individually hybridized with a panel of DNAs from
fifteen different human-hamster somatic cell hybrids to substantiate
this assignment. The hybrids used for this panel were chosen to
provide a unique segregation pattern for each of the autosomes.
The hybridization signal of both DNAs was completely concordant
with the segregation of the origin of the two DNA segments.

The regionally localized X chromosome DNAs described herein, as
well as those isolated by other investigative groups (39,41-43,66-71)
with different methods such as the fluorescence activated chromosome sorting technique described at this conference by Drs. Latt
and Kunkel (42,43), and by Dr. Williamson (41), will permit
direct studies of the structure and the function of the X chromosome,
both in normal individuals and in those with X-linked disorders
such as Duchenne muscular dystrophy.

DR. ROBERT WILLIAMSON: As stated by the previous speakers,
it is clearly better, if possible, to start with clones from only
                              the X chromosome rather than the
Sorting Human X Chromosomes   entire human or mouse genome.
                              Therefore, about three years ago we
began to consider the possibility of sorting chromosomes with
Bryan Young (Beatson Institute, Glasgow). Kay Davies is now carrying
forward this work with X-specific probes in our laboratory (41).
In that paper we describe the preparation of our X chromosome library,
which contains 50,000 clones and should contain all EcoRI fragments
between 4 and 14 kb. Using the FACS-II sorter, chromosome-specific
libraries for most human chromosomes can be prepared.

In one day we were able to sort over one million X chromosomes,
given 500 ng of DNA, using ethidium bromide staining and no special
laser equipment.

It would be optimal to prepare libraries using partial rather
than complete digests, but in this case we wanted a set of clones
as random markers more than we needed a complete and overlapping
library. The X chromosome preparation is contaminated to a small
extent with chromosomes 7 and 8. The clones were first screened
with labeled total human DNA to eliminate clones with a high level
of repetitive sequence, and then were tested for usefulness by
Southern blots to restricted human DNA. Each clone that is useful
(approximately 5 percent of the total) was also screened to ensure
it is human X-specific and contains a sequence of between 5 and 10 kb
so that the autoradiographs develop quickly.

I was very impressed with the lack of background in the blots
from the Boston group. We often find there is what may be a repetitive element in some of our 5-10 kb clones. We do not hybridize
under high stringency and often get a background smear.

Some of our clones have now been localized to regions of the
X chromosome using rodent-human hybrid cells containing only a
portion of the human X (72). In total, over 90 percent of the
clones which can be assigned to a specific human chromosome are X-
specific. The few that have been regionally localized appear to
be randomly distributed along the chromosome.

We also find that most of the clones we have studied show
polymorphisms with at least one person in our panel, but of the ten
or so that have been characterized, only two are frequent enough to
use in patient studies so far. It is not useful to have polymorphisms
in one person out of a hundred and then attempt to study DMD families.
We have shown that the polymorphisms are inherited in a Mendelian
fashion, as expected, but we have no confirmed linkage to DMD at this
time for our polymorphic probes.

We also use library-library screening developed by Julian Crampton
in our research. We feel that this will help in the identification of
differences between individuals. We
cDNA Libraries, Technical   are screening cDNA libraries with
Problems   chromosome-specific genomic libraries
after devising strategies to eliminate
hybridization of repetitive elements (73,74). Unfortunately, we
have not yet been able to screen our muscle cDNA library with the
X chromosome library.

DR. PATERSON: In all of these things with cDNA libraries, don't
you ever worry about the fact that you are losing your more complex
class of information when you are doing this kind of screening, and
all you are looking at is your very abundant material?

DR. WILLIAMSON: We have published much more about the characteristics of our cDNA libraries, than our genomic libraries. The
three papers in NUCLEIC ACIDS RESEARCH discuss the ways of keeping
cDNA libraries constant and stable. The most important point is not
to grow a large number of clones together in medium, in such a way
that they compete with one another in growth.

DR. PATERSON: How do you assure yourself that your cDNA
library is representative of all the genetic information that
you have?

DR. WILLIAMSON: If you hybridize labeled cDNA against the
total pooled library, it shows the same kinetics but displaced by
the expected amount. For this purpose it doesn't matter as it would
for a set of cDNA clones derived from the X and expressed in lymphocytes, in muscle or other tissue cells.

DR. KUNKEL: I might be able to help a bit. If, as Dr. Williamson has pointed out, the cDNA library is propagated as colonies and
not in liquid culture, competition of individual bacterial clones is
kept at a minimum. The resultant colonies should contain insert
sequences which reflect the original RNA complexity. Dr. Kurnit
and I have maintained the libraries of cDNA clones in nitrocellulose
discs as single colonies. All pools were made with isolated DNA
derived from single colony liquid growth. These pools should
resemble the sequence complexity of the original RNA used in the
original cDNA library construction. A pool of 100 was chosen because
of previous defective problems we had in screening with pools composed
of more than 100 cDNA clones.

DR. WILLIAMSON: There are two separate problems, one of the
detection and one of losing sequences during isolation.

DR. KUNKEL: Yes, I agree, but loss is much less important than
detection. Once a cDNA clone has been detected as derived from the
X chromosome then later studies with RNA can determine its relative
abundance.

DR. WILLIAMSON: It is possible to select sequences present in
lower amounts if one uses larger libraries.

DR. KUNKEL: In our hands the signal with a probe derived from
100 cDNA clones is very faint. If we use a higher complexity cDNA
probe the signal to noise ratio is so low that we are unable to
detect a true positive plaque. For this reason we have arbitrarily
chosen a pool of 100 cDNA clones as the highest complexity we can
deal with. We might not be obtaining an optimal phage amplification
in our original plate, but this is dependent upon the number of
phage being plated. What we are both trying to accomplish is to
identify in our libraries of X chromosome DNA unique sequences which
are transcribed into RNA in particular cell types. Probably both
methods will yield such sequences. Bias for abundant sequences can
be sorted out at a later date.

DR. WILLIAMSON: We would like to compare libraries from DMD and
normal muscle. We find differences between normal individuals in
expressed sequences that we do not understand. At least in lymphocytes, there are sequences which do not code for immunoglobulin and
which differ between individuals. This leads us to be cautious in
interpreting results based on differences between individual-specific
cDNA libraries.

DR. HOUSMAN: If I understand the thrust of your comment, you are saying that what if, for example, the Duchenne mRNA is present in a very small number of copies per cell. Let us say there are only ten copies per cell which would mean that there would be perhaps one Duchenne mRNA for every 10,000 mRNAs in the cell. You will pick up clones coded in that region containing the Duchenne gene if your cDNA library is sufficiently complex to represent every mRNA in the cell at least once.

DR. WILLIAMSON: There is no difficulty in obtaining a cDNA library with sufficient complexity. We find our biggest problem is eliminating repetitive elements in library-library screening. For Southern blots, mixtures of probes can create problems purely because of the low concentration of one probe within the mixture.

DR. WHITE: Reproducible patterns may not be achieved because of the inherent variability in hybridization conditions and the low concentration of certain DNA sequences in either the probes or blotted restriction fragments.

DR. LATT: I have a question about this decision whether you screen the plasmids with the phage or phage with the plasmids. First of all, about the amount of material. Is it true that you can get more DNA from the phage plaque then from a plasmid colony? In other words, is a Benton-Davis intrinsically more sensitive than a Grunstein-Hogness with the same amount of counts on a single copy probe?

DR. KUNKEL: Data from Meselson's lab indicate that the use of chloramphenicol amplified plasmids is more sensitive than the Benton-Davis procedure.

DR. LATT: If you use the T4 polymerase labeling method the relative size of insert and vector is irrelevant. If you nick translate plasmids to look at phage, then most of your counts are in your probe. If you are labeling phage to look at plasmids, most of your counts are in λ arms, and it gets to be a pretty hot experiment.

DR. KUNKEL: How did you label the pools of plasmids? By nick translation? We worked with a T4 DNA polymerase labeling procedure which relies on chewback and fill-in. This allows you to label only the insert and not the plasmid vector. It is more difficult to do this procedure with phage DNA samples.

DR. GUSELLA: We haven't done the direct comparison of Benton-Davis to Grunstein-Hogness. We have, however, done the reverse experiment to that performed by Dr. Williamson and have obtained similar results. One in 10^4 counts has to be in specific probe to detect homologous DNA in a particular phage plaque.

DR. HAUSCHKA: If there are about 2×10^5 kb in the X chromosome, and you figure about 1/3 of which are in the short arm where
the gene may be, and since available probes contain about six or
so kb, how many probes and experiments will it take to find the
putative Duchenne gene?

DR. WILLIAMSON: Using linkage analysis, as proposed by
Solomon, Bodmer and Botstein et al (36,75) it should not be necessary
to use large numbers of clones. About twenty, randomly distributed
clones would provide sufficient linkage for all X chromosome markers.
Once a linkage between a clone and the phenotype (in this case DMD)
is established, one could narrow down to that region. It might
be possible to isolate all the cDNA clones transcribed from a small
region and get down to ten or twenty expressed genes, each a can-
didate for the mutant gene.

DR. SINISCALCO: You seem to subscribe to the belief that
with 10-15 probes you can find the linkage along the chromosome.
Are you skeptical about prenatal diagnosis on the basis of linkage
when you are a few thousand kb pairs away from the gene? You need
to find the variation inside the gene and not nearby.

DR. WILLIAMSON: I am doubtful about using linkage for prenatal diagnosis for two reasons. The first is that we do not know
anything about the distribution of recombination along the chromosome, or the extent to which it is sequence-dependent and region-
dependent. We really need to know more about the frequency of
recombination before we can go to patients and begin to talk about
offering prenatal diagnosis with a probe, say, 1000 kb away. In
practice, the furthest anyone has been from a defective gene in
prenatal diagnosis is the work that Peter Little in our lab did
with thalassemia. He used a sequence 25 kb distance from the
β-globin gene. This is much closer than one can come using the
approach that I outlined, at least at first.

The second reason is a practical one. If you are using a
linkage analysis, you must have the right family. You must have
a complete family diagnosis, to exclude nonpaternity, and do all
the sample collection in a week or two as well. Our experience
is that compared to the newer approach for sickle cell anemia,
using an enzyme that recognizes the mutant site itself (76,77),
the situation is markedly less satisfactory using linkage.

DR. SINISCALCO: That is all well and good when you know
what your defect is.

DR. SCHWARTZ: I would like to offer an alternative of rapidly
screening the X chromosome. What if there were some specific X-
linked repeated sequences that are randomly dispersed throughout
the X chromosome? You do genomic blots, frequently cutting enzymes,

and look at the distribution of repeats. You would rapidly detect
any deletion sequences. In other words, you could rapidly screen
200 kb of DNA on a single blot. Why can't you do that and see
whether there is deletion? We can do that for our actin gene that
has repeated sequences 5' tandem to it, and we see reproducible
restriction patterns in checking genomic DNA.

    DR. HOUSMAN: We have used this type of analysis on chromosome
11 to get a rather detailed fine structure map. We start off with
a hybrid cell with only the human chromosome 11. The problem is
that when Gail Bruns and I looked through the X chromosome pieces
for some repeat sequence that wasn't of the normal type, we found
the interspersed repeats were exactly the same as on any other
chromosome. There are repeat sequences which are on the X and either
not anywhere else or primarily on the X. They are interspersed
throughout the X and also particular to the X. That is the problem.

    DR. FRANCKE: Has anyone been successful in making a cDNA
library from human striated muscle?

    DR. KEDES: Peter Gunning and Phyllis Ponte in my lab recently
made a cDNA library from an amputated leg muscle. About 8000 plaques
were selected for full length cDNAs.

    DR. HECHT: Dr. Williamson, how does one get chorionic villi
in the first trimester?

    DR. WILLIAMSON: The chorion is much larger than the fetus at
eight weeks. The obstetricians have isolated villi transcervically
using a thin flexible catheter; they position it against the chorion
using real time ultrasound and obtain one or two villi by suction.
There is a recent series from Moscow claiming no fetal mortality
or morbidity (78) and this will now have to be confirmed at other
centers. Our own series was done on women coming for elective
termination.

    Our experience is that each villus gives 30-40 micrograms of
DNA, enough for several Southern blots. There are several thousand
villi in the chorion.

    DR. KEDES: Could the MDA take on the chore of keeping records
of available libraries and updating and publishing that list?
These libraries could be collected
and distributed in some manner. While
An Available Roster of   I can hardly argue that the technology
DNA Libraries   isn't turning over very rapidly, it
would be very simple to work out a kind of library that one might
want for chromosome assignments from genes that are being cloned
in literally hundreds of laboratories around the world. There seems

to be few laboratories that are interested in constructing these
libraries or making the somatic cell hybrids. I think rather than
having the people try to find out from the literature or by hearsay,
a roster that is updated would be a good idea.

DR. WILLIAMSON: I support the idea in principle, but I fear
the field may be moving too fast. By the time the MDA organized
the lists, the technology would have changed. I am not objecting,
I am just saying that is a tendency to overestimate the usefulness
of this type of exercise when the technology is changing so quickly.

DR. LATT: I would not be surprised if NIH did not do something
like this equivalent to the cell repository. In other words, I
think you might find that the MDA would end up duplicating what
was being done on a larger scale by a larger organization. As Dr.
Kunkel mentioned, anyone who publishes in CELL, for example, does
that with this assumption that the clones are then generally available
I think that is generally the way it should be done.

DR. HOUSMAN: There are several issues that could be talked
about here that have come from our work on Huntington's disease.
It turns out that getting the DNAs
from appropriate sibships is quite
difficult, at least in Huntington's.
Exchange of Information
on Pedigrees
Further, the amount of information
you get by adding the LOD scores from several pedigrees is more
than that from pedigrees that are collected independently. I
think what MDA could expedite is the exchange of information on
pedigrees so that people are not approached twice because it can
be confusing for families if not devastating to be subjected to
more than one approach. The Huntington's volunteers have been
crucial to actually getting the job done properly.

CHAPTER 4

MAPPING THE X CHROMOSOME

DR. UTA FRANCKE: Before moving on to the mapping of the X chromosome, I would like to review briefly with you the background of and progress in human gene mapping. Figure 35 shows the information on human genetic traits that has been accumulated and compiled by Dr. Victor McKusick in his catalogs. The curve depicting the chromosomal assignments of autosomal traits shows a steep rise in the early 1970s due to the advent of somatic cell genetic techniques, and then flattens a little bit because most genes that can easily be mapped in somatic cell hybrids had been mapped. More recently, the slope increased again due to new technology providing new genetic markers such as DNA segments and antigens defined by mono- clonal antibodies. There are now almost 300 loci mapped to specific autosomes, and slightly more than 100 on the X. This information has been reviewed and updated periodically at six international workshops on human gene mapping (79-84).

The prerequisites of chromosome mapping using somatic cell hybrids are the following:

1.   Hybrids have to be constructed in such a way that human chromosomes are preferentially lost. Hybrid clones that contain stable subsets of human chromosomes must be available.

2.   All human chromosomes must be identifiable and distin- guishable from the chromosomes of the non-human parental cell line.

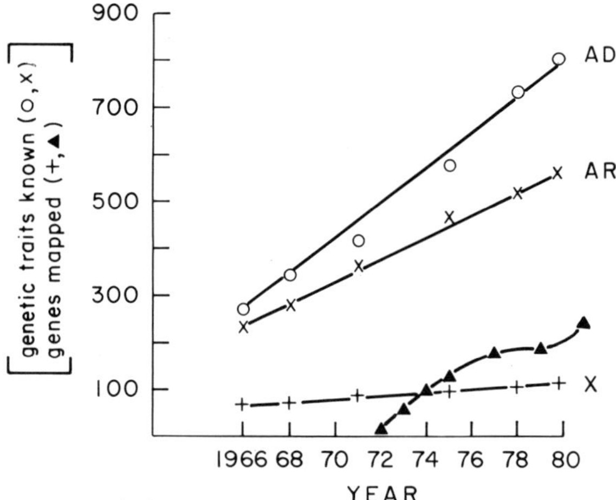

Figure 35: Growth and development of the human
gene map. Data on confirmed autosomal dominant
(AD), autosomal recessive (AR) and X-linked traits
are from reference 107 and previous editions. The
numbers of autosomal traits assigned to specific
chromosomes are based on references 79-84. The
figure is modified from Francke U: Amer J Pathol
101:S41-S51, 1980.

Trypsin-Giemsa or bromodeoxyuridine-acridine orange banding methods are used for chromosome identification and alkaline Giemsa staining for the detection of interspecies rearrangements.

3.   Human gene products in the hybrid have to be detectable. The classical one-dimensional electrophoresis method combined with histochemical or activity staining has led to the assignment of most of the loci for constitutive enzymes to their respective chromosomes. O'Farrell type two-dimensional protein electrophoresis allows one to resolve a very large number of polypeptides. Many of these are distinguishable from rodent proteins and can be assigned to a respective chromosome by two dimensional analysis of hybrid cell extracts (85,86).

A human gene product can also be identified immunologically. Antibodies are available that do not cross-react with the homologous rodent gene products. More recently, hybridoma derived monoclonal antibodies have been extremely useful in studies of complex enzyme systems, for example, phosphofructokinase (PFK), which is coded for by gene loci on three different chromosomes. The human and rodent PFK loci present in hybrid cells produce PFK subunits that combine freely into tetrameric molecules. In this way it is possible that any one hybrid may contain more than 100 isozymic species. These are not resolvable by electrophoretic or chromatographic techniques. With monoclonal antibodies that are specific for certain human PFK subunits, Vora and colleagues have been able to map PFK gene loci (87,88). Monoclonal antibodies are also available against chromosome specific cell surface antigens of known and unknown function (89).

4.   Lastly, cloned human DNA sequences can be localized to chromosomal sites in two ways: by Southern blotting of hybrid cell DNA, and by _in situ_ hybridization directly to human metaphase chromosomes.

Hybrids made with cells from humans who have chromosome rearrangements are useful for regional gene localization. Cells from patients with chromosome imbalances, such as duplications or deletions, are also used for mapping, for example, in gene dosage studies. Individuals with partial trisomies may have elevated levels of a certain gene product or they may express three parental alleles. Individuals with chromosome deletions that include an enzyme locus may have reduced levels of activity. Gene dosage studies have provided many regional assignment data.

The mapped autosomal genes fall into different categories (Table 6). The majority are constitutive enzyme loci expressed in all cells. There are 15 DNA segments

The Current Status of the Autosomal Gene Map

(single copy sequences of unknown function) that are now mapped. Few of them have yet been shown to be polymorphic. The 17 autosomal disease loci have been mapped by linkage to a marker locus (Table 7). For about half of the autosomal loci some regional information is available, and approximately 30 of them have been mapped to a single chromosome band.

About 115 genes are known on the human X chromosome. Most of them are phenotypic traits showing X-linked inheritance (52). Somatic cell genetics has helped to confirm assignments to the X in some cases.

The Map of the Human X Chromosome

Table 8 lists the disorders that have been mapped to the X for which there is somatic cell genetic confirmation that the structural locus (or the mutant gene) is on the X. This has been especially important in the case of testicular feminization (TFM). TFM is associated with a defect in the testosterone receptor. Affected individuals are 46,XY phenotypic females who do not reproduce. In order to prove X linkage, one would have to demonstrate that there is no male to male transmission. However, since the affected males do not reproduce, there is no formal way to distinguish X-linked inheritance from autosomal dominant inheritance with sex-limited expression. The same argument holds true for Duchenne muscular dystrophy (DMD). There is no direct evidence for a structural gene for DMD on the X. The only evidence comes from the females with structural abnormalities of the X chromosome who clinically present with DMD.

Figure 36 illustrates the current regional assignments of loci on the X chromosome. The ideogram represents the 850 band stage of high-resolution G-bands consistent with the International System for Cytogenetic Nomenclature (ISCN, 1981) (90). A total of 38 distinct bands (black, white or gray) are recognizable at this stage. Assignments of loci to cytologically defined regions of the X are based on studies of rearranged X chromosomes in somatic cell hybrids or in patients (under Cytology).

Family studies have established two independent linkage groups on distal Xp [loci linked to the Xg(a) blood group] and on the distal Xq [loci linked to G6PD] (under Linkage).

How does the linkage map (in centimorgans) compare with the cytological map in terms of microns and kilobases? The estimates that are often quoted are based on the following assumptions: 1) Based on the weight of DNA per cell nucleus, it is estimated that there are about 3×10^9 basepairs per haploid human genome.

TABLE 6

Nature and Function of Autosomal Genes Mapped (1981) (N = 271)

Constitutive Enzymes	115	Ribosomal RNA Genes	6
Serum Proteins	23	DNA Segments	15
Proteins with Specialized Function (including hormones)	22	Virus Sensitivity (Host Functions Required)	9
Excretory Proteins	3	Virus Integration Sites	7
Blood Group Antigens	14	Viral Chromosome Modification Sites	4
Other Cell Surface Antigens or Membrane Proteins	25	Autosomal Disorders (Linked)	17
		Others	11

TABLE 7

Autosomal Disorders Mapped by Linkage to Marker Gene (1981)

Chromosome	Marker	Disorder	McKusick #	Symbol
1p	Rh	Elliptocytosis	13050	EL1
1q	Fy	Charcot-Marie-Tooth	11820	CMT1
1q	Fy	Antithrombin-III-deficiency	10730	AT3
1q	Fy	Zonular Cataract	11620	CAE
2p	ACP1	Aniridia	10620	AN-1
4	GC	Analbuminemia	20530	ALB
4	GC	Dentinogenesis imperfecta I	12549	DGI
4q	MNS	Sclerotylosis	18160	TYS
6p	HLA	Adrenal hyperplasia III (21-hydroxylase deficiency)	20191	AH3
6p	HLA	Spinocerebellar ataxia (OPCAI)	16440	SCA1
6p	HLA	Hemochromatosis	14160	HFE
6p	HLA	C2-deficiency	21700	C2
6p	HLA	C4-deficiency	12080	C4
6p	HLA	Hypertrophic cardiomyopathy	19260	
9q	ABO/AK1	Nail-Patella syndrome	16120	NPa
16q	HP	Norum disease (lecithin-cholesterol-acyltransferase deficiency)	24590	LCAT

TABLE 8

X-Linked Disorders with Somatic Cell Genetic Evidence for Localization of the Mutant Gene on the X Chromosome

DISORDER	McKusick #	STRUCTURAL GENE	SYMBOL
HEMOLYTIC ANEMIA	30590	GLUCOSE-6-PHOSPHATE DEHYDROGENASE	G6PD
HEMOLYTIC ANEMIA	31180	PHOSPHOGLYCERATE KINASE	PGK
GOUT WITH ↑ PRPP	31185	PHOSPHORIBOSYL PYROPHOSPHATE SYNTHETASE	PRPS
LESCH-NYHAN DISEASE	30800	HYPOXANTHINE PHOSPHORIBOSYLTRANSFERASE	HPRT
HYPERAMMONEMIA I	31125	ORNITHINE CARBAMOYLTRANSFERASE	OCT
MUCOPOLYSACCHARIDOSIS II	30990	SULFOIDURONATE SULFATASE	SS
ICHTHYOSIS	30810	STEROID SULFATASE	STS
GLYCOGEN STORAGE DISEASE VIII	30600	PHOSPHORYLASE KINASE	PHK
TESTICULAR FEMINIZATION	31370	DIHYDROTESTOSTERONE RECEPTOR	TFM
ADRENOLEUKODYSTROPHY	30010	C26/C22 FATTY ACID RATIO INCREASED	ALD
FABRY DISEASE	30150	α-GALACTOSIDASE A	GALA

TABLE 9

Females with DMD and de novo X/Autosome Translocations

Reference	Year	Translocation	Late replication $[X^n/X^T]$	Phenotype
Verellen (17)	1977	t(X;21)(p21;p12)	L 40/0 F 10/0	DMD (biopsy)
Canki (18)	1979	t(X;3)(p21;q13)	L 54/4	DMD (CPK, EMG, biopsy) MR, dysmorphic
Lindenbaum (19)	1979	t(X;1)(inv p1106 p2107;p3400)	L 75/0	DMD (CPK, EMG, biopsy, histolog. and histochem.)
Greenstein (20)	1980	t(X;11)(p2105;q13)	L $X^n > X^T$	DMD (CPK, aldolase, EMG, biopsy histolog.)
Jacobs (21)	1981	t(X;5)(p21;q35)	L 576/10	DMD (CPK, EMG, biopsy histolog.)
Emanuel (22)	1981	t(X;9)(p21;p22)	L 84/2	DMD (CPK, EMG, biopsy)
Vianna-Morgante (23)	1981	t(X;6)(p21;q21)	NG	DMD

X^n = normal X chromosome; X^T = X involved in translocation
L = lymphocytes;   F = fibroblasts;   NG = not given

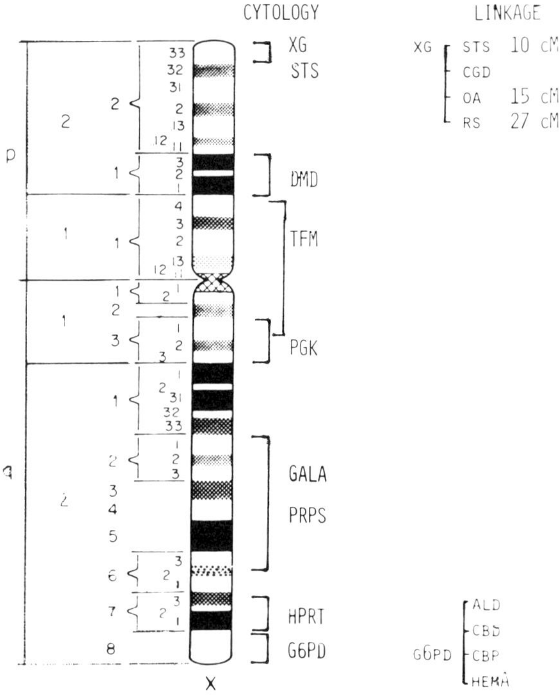

Figure 36:   Regional assignments of gene loci on the
human X chromosome as of 1981.   The ideogram of the
X is from Francke U: Cytogenet Cell Genet 31:25, 1981.
The map is limited to disease loci and loci of known
functions.   DNA segments and cell surface antigens
of unknown function have not been included.   The
brackets under "cytology" indicate gene localizations
on the physical map as determined by studies in somatic
cell hybrids and in patients with chromosome rearrangements.   Under "linkage" the arrangement of the loci
does not reflect the linear order on the linkage map,
which is not clearly established.   The distances are
given in centimorgans.   Linkage of the loci in the
"G6PD cluster" is very tight, with no or very low
frequency of recombination.   Abbreviations: XG=Xg(a)
blood group; STS=steroid sulfatase (ichthyosis); CGD=
chronic granulomatous disease; OA=optic atrophy; RS=
retinoschisis; DMD=Duchenne muscular dystrophy; TFM=
testicular feminization syndrome; PGK=phosphoglycero-
kinase; GALA=α-galactosidase A; PRPS=phosphoribosyl
pyrophosphate synthetase; HPRT=hypoxanthine phosphoribosyltransferase; G6PD=glucose-6-phosphate dehydrogenase; ALD=adrenoleucodystrophy; CBD=colorblindness,
deutan; CBP=colorblindness, protan; HEMA=hemophilia
A.

2)   The length of the genetic map is derived from counts of chiasmata
in male meiosis.   Average counts are around 50 on autosomal bivalents.
Since each chiasma has to be separated from the next one by at least
50 centimorgans (cM), that gives you 2,500 cM for the autosomal map
and roughly 3,000 cM (or 30 morgans) for the total linkage map.
Relating cM to base pairs, this calculation provides an average of
10^6 base pairs per cM.

I consider this estimate to be highly inaccurate for several
reasons.   It assumes that there is uniform distribution of crossovers throughout the human genome, and that is certainly not correct.
It is further based on the male chiasma counts that differ from
female recombination frequencies.   Let us concentrate on the X chromosome.   The cytological length of the X equals about six percent of
the total haploid length.   That would equal about 2×10^8 base pairs
which would correspond to about 200 cM.   This estimate has been used
to calculate how many restriction fragment length polymorphisms are
needed to cover the entire X.   I believe that this is a serious
underestimate.   The calculation is based on chiasma counts in the
male, but recombination between X chromosomes occurs only in female
meiosis.   From comparing linkage distances on the male and female
maps, it is well established that recombination is much higher in
female meiosis.   The best data available involve loci on the short
arm of chromosome 1 (91).   For many pairs of loci on 1p, male recombination fractions range from 9 to 17 percent, while there is no
measurable linkage in females.   There is no easy and accepted formula
that would allow you to convert male map units to female map units.
The relationship appears to differ for different loci.   Chances are
that recombination frequencies differ not only between the sexes but
also according to the region of the chromosome.   Therefore, the
question of how many kb equal how many cM does not have a uniform
answer.

With respect to the X map, the loci for G6PD and HPRT map to
neighboring bands on the distal long arm (Figure 36).   When I was
in Dr. Nyhan's laboratory in San Diego, we studied offspring of
women who were double heterozygotes for G6PD A and B and for HPRT
deficiency (Lesch-Nyhan disease).   There were five recombinants out
of thirteen informative offspring (92).   Taking into account the
possibility that some of the non-recombinants may in fact be the
result of double crossovers, these data indicate that G6PD and HPRT
are not linked (by Kosambi formula the linkage distance is 51 cM)
(92).

On Xp, the loci for the Xg(a) blood group (XG) and for steroid
sulfatase (STS; X-linked ichthyosis) have both been mapped to the
most distal subband, while the recombination fraction between them
is about 10 percent (84).

More data need to be collected on linkage between other loci
on the X that have also been mapped cytologically, before one can
estimate the total genetic length of the X realistically. While
it is possible that recombination frequencies are higher towards the
ends of chromosome arms, the available data suggest that the genetic
length of the X is much greater than 200 cM. If it were on the order
of 500 cM, the short arm, representing 2/5 of the cytological length
at the 850 band stage, would contain 200 cM. This could explain
why the DMD locus, provisionally assigned to band Xp21 in the middle
of Xp, has been found to be unlinked to both G6PD and XG (data
summarized in reference 93).

The evidence for the localization of DMD on Xp21 is based on
cytogenetic studies of females with classical DMD. In females expressing an X-linked recessive disease,
Evidence for Mapping of the   a number of possible explanations have
DMD Locus to Band Xp21   to be considered: 1) An affected
                                father and a heterozygous mother can
produce a homozygous affected daughter. 2) With father affected,
the maternally derived X could carry a new mutation. 3) The mother
could be heterozygous and the paternally derived X could carry a new
mutation. In these three possibilities the affected female is homozygous for the mutant allele. The first two possibilities are ruled
out for DMD because affected males do not reproduce. 4) Hemizygous
expression of an X-linked recessive trait in a female can be associated
with a chromosome abnormality: 45,XO or non-random X inactivation due
to structural rearrangement. For example, a sizeable deletion of
part of one X chromosome leads to the binding of preferential or
exclusive inactivation of the abnormal X are presumably due to
selection in early embryonic life. In this case, a mutant gene on
the normal X would be expressed clinically. X-autosome translocations
can lead to preferential inactivation of the normal X if the break
is within the X chromosome but not close to the telomere regions.
A mutation present on the X that is involved in the translocation
would be expressed. 5) The last and most intriguing possibility
suggests that an apparently balanced X-autosome translocation can
render a gene at or near the X chromosomal breakpoint non-functional,
thus leading to expression of an X-linked mutant phenotype. This
seems to be the case in seven reported females with classical DMD
and a translocation involving a break in band Xp21. The available
information consists of three full reports and four cases reported
in abstract form (Table 9) (94-100).

The first question to address is whether these patients really
have DMD, or some other form of muscular dystrophy. In terms of
Dr. Brooke's criteria, all of those for whom information is available
had onset of symptoms at less than 5 years of age. Two patients were
8 years old, one of them wheelchair-bound (92) and the other one
hardly able to walk (98). They had proximal weakness and pseudo-
hypertrophy. All patients had greatly elevated CK levels; EMG findings were characteristic of DMD. No sensory involvement has been
reported.

In five cases muscle biopsies were studied histologically, and in
one of them histochemically as well. The findings were diagnostic
for DMD. One of the patients also had multiple dysmorphic features
and mental retardation (95). In this case, part of chromosome 3
had been exchanged with the short arm region of the X. In 4 out of
58 cells studied the translocated X was late replicating with apparent
spreading of the late replication into the translocated autosomal
segment. Inactivation of autosomal material could explain the dysmorphic features and mental retardation in this individual.

Chromosome studies in the 7 girls with DMD revealed an X-autosome translocation involving a break in band Xp21 and breaks in seven
different autosomes (chromosomes 1, 3, 5, 6, 9, 11 and 21 have been
involved). The translocation was de novo in all instances, with the
parents having normal chromosomes. Chromosome replication studies
indicated preferential or exclusive late replication (inactivation)
of the normal X chromosome.

Family histories are significant for the absence of other
individuals affected with DMD. There are five reported male siblings
altogether, none of which is affected. In the earliest reports,
two of the mothers were said to have elevated CK levels (94,95).
With respect to possible origin of the mutant X chromosome it is of
interest that two of the fathers were of advanced age: 40 and 39
years old (95,98).

If one assumes that these are all biopsy proven cases of classical DMD, the reported findings can be explained in two ways. The
first involves the presence of a mutant DMD allele which is expressed
clinically because of non-random X-inactivation. Under this hypothesis, the normal allele on the normal X chromosome is inactivated,
and the mutant DMD allele could be located anywhere on the X involved
in the translocation. The DMD allele could be inherited or the
result of a new mutation. This explanation was considered plausible
when the first two cases were identified (94,95). The presence of
seven unrelated such cases makes it much less likely. Aside from
the two mothers who had possibly elevated CK levels, there is no
evidence that the DMD mutation was inherited. Therefore, it is
very likely that the DMD mutations were de novo. The translocations
were de novo in all cases as well. Taking the mutation rate for
DMD at 7×10^{-5} and the mutation rate for a new structual abnormality of this nature in the range of 10^{-4}, the probability of these
two mutations occurring together would be in the range of 10^{-8} to
10^{-9}. Furthermore, numerous X-autosome translocations have been
reported that were associated with non-random inactivation of the
normal X chromosome but with breakpoints on the X in sites other
than Xp21. None of these females had DMD. The specificity of the
breakpoint in Xp21 in females with DMD provides strong evidence that
these cases are not co-occurrences of two unrelated rare events

involving the same X chromosome, one being a structural rearrangement
(translocation), and the other being a change at the DNA level.

DR. EPSTEIN: How many cases of translocations in that region
are known that do not have dystrophy?

DR. FRANCKE: None that I am aware of. Most of the known X-
autosome translocations involve the long arm, or the short arm more
distally, or the short arm near the centromere. I am not aware of
a translocation involving Xp21 that is associated with a normal
phenotype.

Could there be an effect of the autosomal genes that have been
placed adjacent to genes on Xp by the translocation event? The
autosomes involved were all different in the seven cases. Therefore,
a "position effect" on an autosomal gene is an unlikely explanation.
What seems to be important is the disruption of DNA sequences in
the Xp21 region that interferes with the functioning of the normal
allele at the DMD "locus." The event may either break the relevant
stretch of DNA directly or remove it from a controlling region that
it has to be in contact with. This hypothesis postulates the location
of the DMD locus in band Xp21.

DR. HECHT: I agree with your preference in terms of explanation.
There was a lot of discussion yesterday about whether muscular
dystrophy involves an abnormal gene product and I think the conclusion
was that there is no evidence for that. If you were to extrapolate
from this, one real possibility is that Duchenne muscular dystrophy
is analogous to certain types of α-thalassemia and it is basically
a deletion disease.

DR. FRANCKE: You mean a deletion disease in that way that
genetic material has been removed and no abnormal product can be
identified, but possibly a lack of a product.

DR. HECHT: That's right. There is lack of either something
that is structural or something that is controlling so that nothing
abnormal is made but something is not made that should be because
what you basically have here by this hypothesis based upon the
seven cases that have so far been reported of X-autosome translocations, they only have two common features, one is the breakpoint
in the X and the other is Duchenne muscular dystrophy. What you
have from this in functional terms is deletion homozygosity or at
least you have got a heterozygosity for a deletion and maybe that
alone is enough to do it.

DR. FRANCKE: It is a functional hemizygosity, because we
know that this locus is inactivated, it is not in the region of
the X that escapes inactivation. The normal allele on the normal X

chromosome would be inactivated, leading to hemizygous expression
of the mutation, which means somatic cells are lacking the respective gene product.

DR. SHAPIRO: If I understood what you said correctly in terms
of the summary, six of the seven females are apparently balanced.
Have all of them been subjected to high resolution banding and how
many normal Duchenne patients have been looked at with high resolution banding?

DR. FRANCKE: By standard cytogenetic analysis, the translocation is apparently balanced in all seven patients. The abnormal
phenotype in one of them is presumably due to low frequency of
inactivation of the translocated X, with spreading of the inactivation
into the autosomal segment (95). There are no reports on high-
resolution banding (HRB) of these patients' chromosomes. Therefore,
it is not clear whether all seven X-autosome translocations involve
the same subband of Xp21. I am not aware of HRB studies in male
DMD patients but Dr. Siniscalco has data on this topic.

DR. SINISCALCO: Concerning the question of whether or not
Duchenne muscular dystrophy might be due to a deletion in the
affected females with the Xp21 translocation, it seems to me that
the pattern of inactivation in these patients (who all regularly
show preferential inactivation of the normal X) is against such a
hypothesis. We have been interested in the question raised by
Dr. Shapiro concerning high resolution banding. During the past
summer, Dr. O.J. Miller at Columbia University and Dr. R. Chaganti
at Sloan-Kettering screened with prophase banding a great deal of
material, too much for any single laboratory, let alone mine. We
studied twelve uniplex families with an individual affected and
nothing else as well as all of the other segregating families. We
found no instance in this material of chromosomal abnormality with
a prophase banding.

DR. HECHT: I am not suggesting that in Duchenne under normal
circumstances one is going to be seeing a cytologically detectable
deletion in this band under the microscope. What I am suggesting
and I think it is in line with what Dr. Francke is saying, is that
this is functional deletion, hemizygosity that is occurring and
that one just happens to have a unique experiment of nature with
these seven children in that one can see a rearrangement at that
point. What the rearrangement may be doing is simply disrupting
transcription, translocation, whatever. It is not necessarily an
actual loss of chromosome material; it may simply be a local
disruption of genetic information as Dr. Francke suggests.

DR. FRANCKE: I agree with you. I think that a deletion that is visible in the microscope in that region would probably have more severe effects than just Duchenne's. At least you would expect some degree of mental retardation.

DR. WILLIAMSON: Dr. Francke, as you said, the classic analysis of the X chromosome in terms of cM comes up with an answer that you need something in the order of 10, 15, 20 probes to cover the whole of the X chromosome for linkage analysis.
I wasn't aware of the argument about the X having a very much larger distance. It is really the comparative male/female map I would like.

When we are talking about cM, the distance on the X can be derived either from the total number of chiasmata in the total meiotic picture or from the recombination distances. On the basis of the data you presented on the female, cM distances mapped by crossover by actual linkage analysis, how many probes do you think would be needed to cover the short arm from XG down to the centromere for a linkage map?

DR. FRANCKE: As I said earlier, the data are too limited to come up with a realistic estimate. If you would take the genetic distance between HPRT and G6PD and translate it to all of the X chromosome, the total genetic length of the X could be as much as 600 cM; there could be 250 cM on the short arm alone. But there is evidence for increased chiasma frequencies towards the ends of chromosome arms for certain autosomes. Therefore, you would expect that the X linkage map is not strictly proportional to the cytological map. At any rate, I believe that the genetic length of the X may be much larger than current estimates in the literature suggest, which would greatly increase the number of restriction fragment length markers you would need to cover the whole X.

DR. WILLIAMSON: It doesn't increase it out of sight though?

DR. CASKEY: We are attempting to understand spontaneous mutants at one locus on the X chromosome, the HPRT locus. We have examined in Chinese hamster cells a single step conversion from HPRT + to - and have found, in two of ten cells analyzed, lack of Southern bands, just a total set missing, but certainly evidence that looks suggestive of a deletion event. Those cells have not been characterized cytogenetically at this point but it certainly sounds suspiciously like deletion maybe a reasonably common spontaneous conversion of this locus. We are just initiating the studies with the Lesch-Nyhan cases. The probe does cross-react well against human. We hope that with a large enough spectrum of mutants we will come to some understanding to whether there are deletion or other type alterations. I think we should be able to answer a lot of detailed questions about this particular locus on the X chromosome using these probes.

DR. FRANCKE: One may have to consider that there could be
something peculiar to Chinese hamster cells in culture. Siminovich
has suggested some years ago that they may be functionally hemizygous at a large number of loci, and Ron Worton has evidence for
the inactivation of certain loci on Chinese hamster autosome in
cultured cells.

DR. HECHT: The discussion has largely centered naturally
around Duchenne, where would you place Becker and maybe Emery-Dreifuss
(101). Where would you put the other kinds of muscular dystrophy
that are thought to be on the X?

DR. FRANCKE: There is weak evidence for linkage of Becker's
muscular dystrophy to the G6PD cluster at a moderate distance. But
the LOD score is not up to the level that would prove linkage.

DR. HECHT: What do you think about the possibility that Becker
may be actually up in the vicinity of Duchenne and may represent the
gene duplication?

DR. FRANCKE: You mean an allelic variant of the same locus?

DR. HECHT: There are two possibilities. One that it is another
allele at the same locus but the other possibility would be that gene
duplication occurred with a slightly divergent evolution afterwards.

DR. FRANCKE: That is an interesting hypothesis. Your guess
is as good as mine.

DR. EPSTEIN: I wonder whether you could offer an explanation
for the difference in recombination frequencies between male and
female? Are there any mechanisms that have been discussed?

DR. FRANCKE: Biologically, the way they both proceed is quite
different. The female meiotic cells are essentially arrested in
the first meiotic division by the time the girl is born and they
will remain in that stage until after fertilization. One would
think they have a lot more opportunity for crossovers just because
of the time factor, while male meiosis proceeds much more rapidly.
There will certainly be other factors that are not well understood
at this time.

DR. LATT: I recognize that the map length and recombination
frequency in females is generally considered to be greater than
in males. Is there really evidence for this throughout the genome
or is it just an overall impression derived from a few specific
examples?

DR. FRANCKE: The 1p is picked because there are more data on
comparing physical and genetic maps in this region, but it is true
for all linkages that have been looked at. It is also true for the
mouse and Drosophila. It is not something peculiar to humans.

DR. LATT: It has been checked on lots of other chromosomes?

DR. FRANCKE: Yes, of course. In human linkage analysis the data
on recombination are calculated separately for males and females and
sex-specific maps are being constructed (93). To quote N. Morton
(82): "The concept of a 'neuterized' map is not useful in a bisexual
species."

DR. HAUSCHKA: I have a question that comes out of the comments
you made about somatic genetic analysis. In these experiments you see
what appears to be a loss of, let us say human chromosomes for which
there is no selective advantage to the hybrid cell and it brings up
the question as to why in the experiments that we have heard about in
human cell lines which have XXXX chromosomes there is retention of all
of those extra X chromosomes and why aren't they lost. Why are those
retained in the lines? To me, I guess it suggests the possibility that
there may be a selective advantage in keeping all of those. They may
be functional at a low level and I wonder if people have thought about
that as some sort of explanation for why the situation in carriers for
Duchenne may vary so much that is there is really no completely inactive X. Have you or other people thought about that problem in
relation to the XXXX human cell line?

DR. FRANCKE: Can I just say something about chromosome segregation first? And then Dr. Shapiro can comment more specifically.

I think you are comparing cattle and fish here, because chromosome segregation in aneuploid cells and in interspecific somatic cell
hybrids are two separate issues. Meiosis in aneuploid cells occurs
normally without preferential elimination of extra chromosomes, although
occasionally chromosomes can be lost or gained at random in lymphocytes
or other cultured cells. But in interspecific cells you have a totally
different situation. Foreign chromosomes are present in a different
environment. Although this system has been exploited for gene mapping
for about ten years, no one has come up with a good explanation why
one species of chromosomes are preferentially lost. The mechanism is
not understood.

DR. SHAPIRO: In regard to that, it is exceedingly difficult to
prepare somatic cell hybrid lines that retain an inactive X, much
more so than it is to randomly retain an autosome, for example.
That may have something to do with the replication time of that X
in the cell cycle, but it may have something to do with other reasons
as well. We have three different somatic cell hybrid lines in which
we have been able to retain the inactive X and to manipulate it

with a variety of rather sensitive methods for expression of various
genes from that so-called inactive X and most of the techniques would
allow detection of as little as two or three percent of gene product
of wild type levels and one does not see expressions, so within those
limits, I think it is true. I would point out that there is some
dlightly contrary evidence in marsupials in which the paternal X is
preferentially the inactive X in essentially all marsupial species
that have been looked at. There is some evidence, however, for low
level of expression of certain gene products from the marsupial X
chromosome. There are segments of the human X, however, which appear
to escape the usual process, at least of inactivation as Dr. Francke
has mentioned. I guess in terms of the well characterized genes that
would be XG and steroid sulphatase. However, I think there is starting
to accumulate some data. Again, maybe we could discuss this a little
bit later, with regard to steroid sulphatase that would suggest with
respect to the failure of inactivation or level of inactivation, for
example, with dosage with regard to steroid sulphatase one doesn't
see the two to one gene dosage in females as compared to males, but
sees something somewhat less and there may be some other explanations
for that. At any rate, I think most people would agree that there is
not really any evidence for expression of gene products from the
inactive X.

    DR. BOYER: An X-linked marker tells us about the lyonization or
X inactivation within the multiple nuclei of striated muscle fibers.

    DR. FRANCKE: There is some evidence from G6PD heterozygotes,
women who are heterozygous for the electrophoretic variants G6PD A
and B. In addition to the A and B forms, they produce an AB hetero-
dimer. In the syncytia there must be nuclei expressing A as well as
B subunits. The A and B subunits combine in the cytoplasm to form
the heterodimers. The quantification of heterodimers gives information with respect to the degree of lyonization.

    DR. JAMES GUSELLA: What I am going to tell you about could
apply directly to what Dr. Francke spoke about and is another method
of obtaining a fine structure map of a
Mapping Autosomes   particular chromosome that doesn't
as a Model   require doing 50 or 60 Southern blots
                           with 50 or 60 single copy probes. What
we have done is to put together a fine structure map of chromosome 11,
in this case with A36FC, a repetitive sequence probe, which hybridizes
to multiple bands on the chromosome. It is a non-Alu repetitive so
it is only repeated approximately 2000 times per genome and on chromosome 11 it will detect 30 to 40 fragments. A few years ago we mapped
the β-globin locus using a panel of cell hybrids which were derived
from one original hybrid (J1) that had only human chromosome 11 on
a CHO background (102). That original hybrid was irradiated and
deletions were caused in the chromosome 11; sub-clones were picked,
each of which had a different region of chromosome 11. Using hybridization of a β-globin probe we mapped the β-globin locus to the region

between ACP-2 and LDH-A (Figure 37). We have used that same panel
of cell hybrids in this study. We have one cell line that appears
cytologically to show a break at q13 just below the centromere (J1-11)
and therefore contains from this region to the terminus of the short
arm. We have a series of cell lines (J1-7, J1-10 and J1-23 respectively)
which shows breaks at p11 (below ACP-2 on the short arm), p12 (between
ACP-2 and LDH-A and p13 above LDH-A) with a variable region of the
short arm. We have another cell line (J1-4b) that appears cytologically to contain only the centromere region. The apparent breakpoints
are q13 and p11.

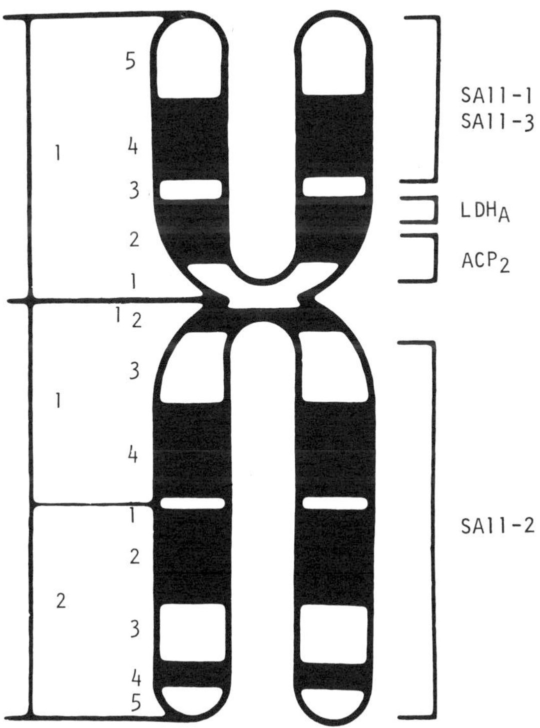

Figure 37: A schematic representation of human
chromosome 11 showing the banding pattern with
its numbering scheme and the approximate loca-
tions of several marker loci.

DNA from each of these cell lines and CHO was digested with EcoRI, run on an agarose gel, transferred to nitrocellulose and hybridized with the A36FC repetitive sequence probe (Figure 38). This sequence was originally cloned from the region immediately 5' to one of the human γ-globin genes located on the short arm of chromosome 11. As a result it hybridizes intensely to a fragment of approximately 7 kb in all the hybrids tested. However the probe also hybridizes less intensely to multiple fragments which appear to occur at many different places on chromosome 11. The majority of the bands in the J1-11 lane represent bands either on the short arm or in the centromere region. The J1-4b lane is interesting because if you look at this hybrid cytologically, it appears to contain only the centromere with little else. If you look at the A36FC pattern, however, this hybrid appears to have a breakpoint on the short arm that is somewhat above the breakpoint of the J1-10 hybrid. It therefore actually contains about half of the short arm based on this hybridization. Thus we have actually been able to map out what part of chromosome 11 is present in a more exact fashion than we could by looking at chromosomes under the microscope. We have now put together a map with a number of these bands using many more such hybrid cell lines and it is quite clear with some of the hybrids that they are certainly not what they appear to be by karyotype analysis. The method described here has detected internal deletions in some of the chromosomes which are not detectable cytologically. You can see that with a probe like this, if you can compare chromosomes directly out of hybrid cell lines you can easily pick up pieces of DNA that are absent from deleted chromosomes and those fragments can be cloned directly using ths repetitive sequence as a probe. There are also of course a number of other uses for such probes. One would be to monitor hybrid cell panels to determine what chromosomes are actually present by the pattern of repetitive-hybridizing fragments rather than by using only the protein markers that are currently available. It might also be useful ultimately when the Duchenne muscular dystrophy locus or some other locus of interest is located on a chromosome to enable one to follow that particular region of chromosome through a series of re-ductions in somatic cell hybrids in an attempt to obtain the smallest amount of human material possible in the hybrid and therefore allow cloning of the region very close to the locus in question. I think this would actually fit together very well with Dr. Francke's hybrids if she has a hybrid with the deleted chromosome and the non-deleted chromosome. Any fragment that appears in one but not the other would likely come from the deleted region.

DR. SINISCALCO: Did you use this probe on the original individual?

DR. GUSELLA: It wasn't used against the original individual; it was used against the original hybrid that contains the entire chromosome and the entire array of bands that is in any of the de-letion hybrids is present in that.

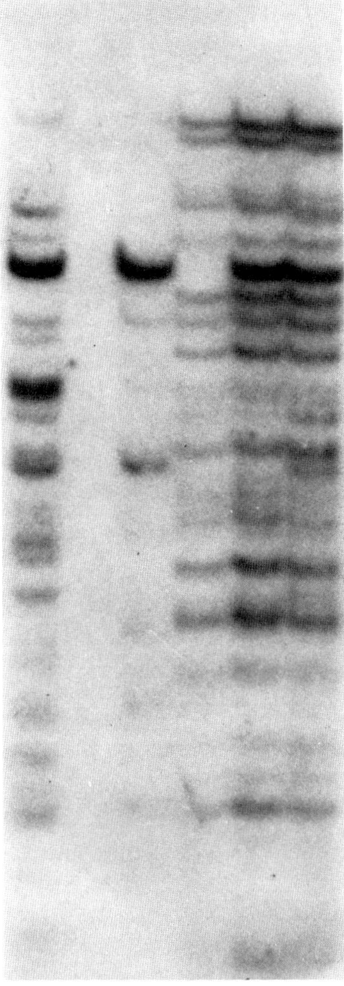

Figure 38: Pattern of hybridization of A36FC repetitive sequence probe to human chromosome 11. 10 μg of DNA from each of the indicated cell lines was digested with EcoRI and the resulting fragments were resolved by agarose gel electrophoresis followed by transfer to nitrocellulose and hybridization to A36FC a human repetitive sequence probe. The portion of chromosome 11 present in each hybrid line is described in the text.

DR. SINISCALCO: Still, even against the hybrid, at this point we don't know whether there is any somatic variation especially when using a probe which gives a multiple pattern. I wonder, since the hybrid is one cell whether the cloning of somatic cells would disclose any of these changes of the RFLP type.

DR. GUSELLA: I will show a blot which is essentially two-dimensional, looking at the entire human genome with two different enzymes. We run one enzyme in the first dimension, another enzyme in the second dimension. You don't pick up an enormous degree of polymorphism using this probe. There may be an occasional difference but by and large most bands are the same. This is just comparing lymphoblast lines so it is not different tissues.

DR. SCHWARTZ: How large is your repeated sequence?

DR. GUSELLA: 2.2 kb.

DR. SCHWARTZ: What is the degree of stringency that you use in your hybridization?

DR. GUSELLA: The most stringent wash is a 1 x SSC, 65°. I should also point out that the CHO lane in Figure 38 shows no cross reaction of hamster DNA with this probe.

DR. DAVIDSON: The band of origin was more intense than the other bands in the blot. Is that a matter of repetition?

DR. GUSELLA: We were trying to figure that out right now and it appears that the repetitive sequence we have got is actually two different repetitives that can be separated and that both of them, therefore, hybridized to that one band. The differential intensity could also be due to different degrees of homology with the cloned sequence. Another point is that this is not the only repetitive sequence you can do this with. You can potentially do it with any non-Alu repetetive. We have isolated a large series of these which we are now characterizing. Hopefully they will have different repetition frequencies and therefore we will have the ability to resolve different amounts of human DNA to an optimal degree.

DR. MARCELLO SINISCALCO*: I have been amazed to realize how much of convergent evolution there has been in the devising of experimental strategies for the molecular mapping of the human genome. For some time, my colleagues and I have been stressing the potentials of highly reduced rodent-human hybrids for the isolation of nucleic acid probes from specific human chromosomes or chromosomal regions (103-106). Our initial efforts were

Screening for Molecular Markers of X-Linked Muscular Dystrophies in Sardinia

*In collaboration with Drs. P. Szabo and G. Davatelis (120).

directed towards the use of a mouse-human cell line (A9/HRBC2) which had essentially retained only the human X chromosome (Figure 39) to attempt the isolation of human X chromosome-specific mRNA. The obvious continuation of this work would have been the preparation of the corresponding cDNA families and, eventually, their cloning in suitable vectors. These plans, however, had to be shelved for several years in view of the DNA moratorium as well as for the difficulties of obtaining adequate support for a research endeavor. After the lifting of the moratorium and the convincing demonstration given by Gusella et al (44), that highly reduced rodent human hybrids were indeed an efficient biological tool for the isolation of chromosome specific DNA sequences and their mapping, we resumed our original plans with a change of strategy inspired by their success. I will summarize here the status of our studies with special reference to their relevance to the genetics and biology of X-linked muscular dystrophies.

With the intent of constructing an array of DNA probes for specific regions of the human X chromosome, we have prepared two DNA libraries in λ phage vectors (Charon 28 and gt7) using DNA from a murine-human somatic cell hybrid, A9/HRBC2 (Figure 39). This cell line

Cloning of Random Human
X Chromosomal Sequences

is a highly reduced hybrid containing only the human X chromosome and a minute fragment of autosome 2. The Charon 28:A9/HRBC2 library, which was prepared by ligating a Sau 3A partial digest of A9/HRBC2 DNA into the Bam HI arms of Charon 28, does not contain inserts as large as we had originally hoped for. For some yet undetermined reason, the bulk if not all of the recombinant phages also contain the dispensable Charon 28 insert fragment, even though only a slight trace of this fragment could be detected in the preparation of arms used to produce the library. However, most if not all of the phages that were generated do contain A9/HRBC2 inserts, the average size being about 4-5 kb. Because of the small insert size, the 7.7 x 10^5 plaques in this library represent only 75 percent of the sequences found in the hybrid cell, not the 90 percent we originally estimated. Since we are only interested in obtaining a number of random, single copy sequences from the human X chromosome, this library is sufficient for our purposes. No problems of this type have been encountered with the gt7:A9/HRBC2 library which appears to contain all EcoRI fragments of A9/HRBC2 DNA between 7-14 kb in length. Both libraries have been screened for the presence of human DNA inserts, and ninety clones from the Charon 28 library have been plaque purified.

To ensure that the DNA libraries contained the expected inserts, a series of screenings was done with total mouse and human DNA as well as several cloned probes for sequences of interest to some of our collaborators. The results are summarized in Table 10.

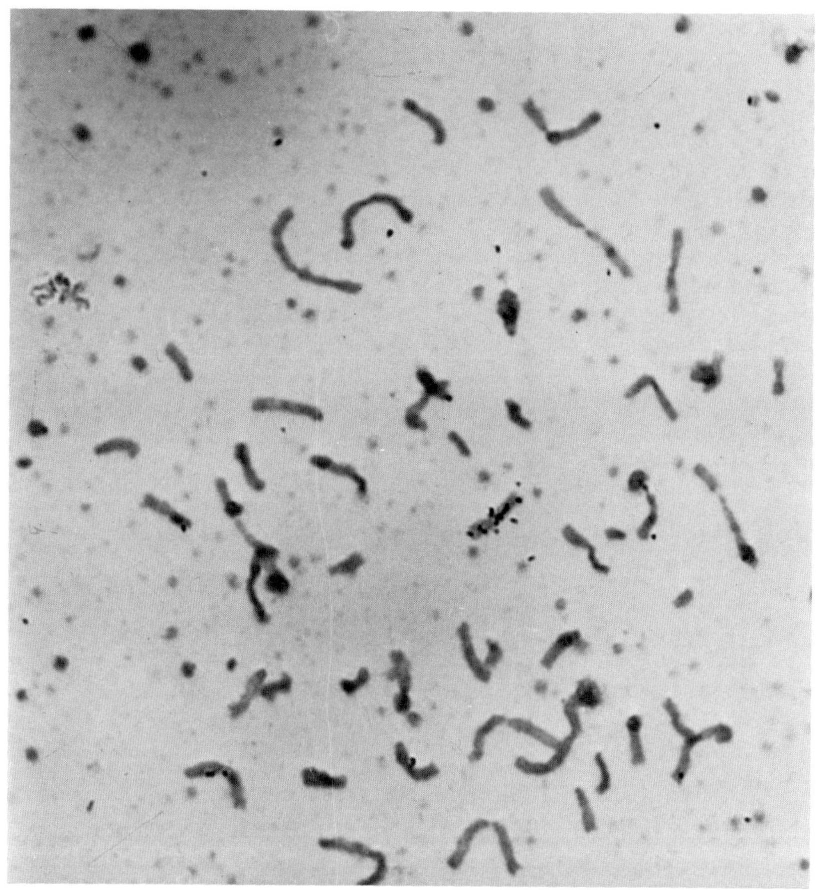

Figure 39: A metaphase of the rodent-human line
HRBC2 which was used for the molecular cloning of
the human X chromosome because cytogenetical and
biochemical data had indicated that this line contains only the human X chromosome and a small fragment of autosome 2 (105). In this metaphase, the
human X chromosome was identified by <u>in situ</u>
hybridization with a ^{125}I-labeled recombinant phage
DNA clone containing a human Alu I sequence which,
under stringent conditions, does not react at all
with the murine chromosomes as shown in the Figure
in which only the human X is labeled. The circumstance is the basis of the selection strategy
reported in Figure 40.

TABLE 10

Summary of screening for DNA recombinant clones with human X
inserts from DNA libraries prepared from a mouse–human hybrid
cell line with only the human X chromosome (A9–HRBC2).

(i)   A9/HRBC2 library in Charon 28 (Sau 3A fragments)

Types of probes	Total # of plaques screened	# of positive plaques	%
^{32}P–total mouse DNA	66	45	70
^{32}P–total human DNA	6.0×10^4	640	1.6
^{32}P–actin cDNA	2.4×10^5	13	.0054
^{32}P–4 kb Bam HI interspersed, repeated mouse DNA	4.0×10^3	310	7.7
^{32}P–0.54 kb Bam HI interspersed, repeated mouse DNA	4.0×10^3	380	9.5

(ii)   A9/HRBC2 library in gt7 (EcoRI fragments)

Types of probes	Total # of plaques screened	# of positive plaques	%
^{32}P–total human DNA	10^4	357	3.6
^{32}P–0.54 kb Bam HI interspersed, repeated mouse DNA	10^4	770	7.7
^{32}P–1.3 kb EcoRI interspersed, repeated mouse DNA	10^4	80	0.8

Clearly both human and mouse inserts are present in both libraries
Some cross-hybridization is expected and was observed for some of
the purified phages, about 20 percent of those purified against
probe total human DNA hybridized to both human and mouse DNA.

Besides the screening with total DNA, a series of cloned interspersed repeated DNA sequences of the mouse and a Chinese hamster
actin cDNA clone were also used to screen these libraries. The
number of positive phages obtained with each of these probes is
within the range expected based on the repetition frequency of each
sequence of interest in the mouse genome. The family of interspersed repeated mouse sequences provides a good example. These
sequences can be seen as bands above the streak of Bam HI and EcoRI
digested mouse DNA by ethidium bromide staining. A number of these
bands (4 kb and 0.54 kb Bam HI fragments and a 1.3 kb EcoRI fragment,
which is part of the 4 kb Bam HI fragment) have been cloned by
Drs. P. Soriano and M. Meunier in Dr. G. Bernardi's laboratory.
We are using these cloned fragments to isolate phages from our
libraries which contain the junctions of the interspersed sequences
with adjacent mouse DNA. As is clear from the table, the number of
clones homologous to these sequences is high, roughly 7-10 percent
of the plaques. This is expected, since these sequences have a
repetition frequency of the order of $1 - 2 \times 10^4$ per genome. The
actin genes (cDNA clone provided by Dr. P. Soriano) which are found
in 10-20 copies per genome are seen in only 0.0054 percent of the
plaques as expected on the basis of the difference in repetition
frequency.

It is interesting to note that in the case of the screening
of the gt7:A9/HRBC2 library with the 1.3 kb EcoRI fragment relatively few positive clones were obtained. This observation is
consistent with the expectation that the 1.3 kb EcoRI fragment
found in most of these repeated sequences would not be cloned,
since we selected 7-14 kb EcoRI fragments to prepare this library.
Those phages which were labeled probably derived from repeated
units which lack one of the two EcoRI sites. Such variants
account for approximately 10 percent of the repeat units (M. Meunier,
personal communication).

These studies have proven that the libraries show the expected
frequency of particular inserts of known repetition frequency.
Consequently, we concluded that we would be able to isolate human
clones of the type we needed by screening with total human DNA.
This type of screening identifies phages which contain repeated
human DNA sequences, either in tandem or interspersed. Those containing interspersed repeated sequences should also contain adjacent
single copy sequences. This expectation has been confirmed for most
of the purified human clones isolated from the Charon 28:A9/HRBC2
library, in that they all appear to contain interspersed repeated
DNA.

The purification of phages containing sequences homologous to total human DNA was accomplished using the scheme shown in Figure 40. About 200 positive plaques of three classes, heavily labeled, moderately labeled and lightly labeled, were transferred by sterile toothpicks to a fresh lawn of bacteria. The lightly labeled phages were included on the assumption that they may contain human sequences which were only moderately repeated. Our subsequent analysis of the purified clones suggested that, although it was valid in some cases, many of these phages are heavily contaminated with total mouse DNA. This suggests that they are derived from the mouse genome but have some homology with repeated human DNA.

The next step in the purification was taking a plug of agar from the "toothpick" plates, suspending it in λ dilution buffer and streaking the phage on a fresh lawn. This was done for about 150 plaques. Those containing human DNA inserts were identified by hybridization and the cycle repeated two or three times, depending on the phage, until all the phages were uniformly labeled. At present, we have ninety plaque purified phages and twenty additional ones which require another round of purification.

Prior to the last round of purification, small scale DNA preparations were made from 35 of these phages. Restriction enzyme digests of these samples showed that most were already pure and that almost all of the others contained only low levels of single addi-tional phage. The bulk of these 35 samples hybridized strongly with human DNA. Eight phages hybridized better with mouse DNA than human DNA; one hybridized with both DNAs but much better with human DNA. Figure 41 shows restriction maps of five of these phages with inserts ranging from 4.6 to 5.4 kb.

Hybridization of total human DNA to Southern blots prepared from gels of EcoRI, Bam HI or EcoRI + Bam HI double digests has allowed us to determine which fragments contain repeated sequences and which contain single copy sequences. As is evident from the maps, a number of fragments containing single copy sequences can be prepared from these phages using only EcoRI and Bam HI. It is likely that others could be produced using different restriction enzymes. For convenient subcloning, we concentrate on enzymes which produce fragments which can be easily subcloned in pBR322 or pBR325. We will subclone all single copy sequences that -- owing to their size and their region of homology along the X chromosome -- promise to be of particular value for the screening of multiallelic and common RFLPs or restriction fragment length polymorphism as the one described by Wyman and White (107).

We are currently evaluating an alternative and perhaps more convenient strategy for the accumulation of few copy recombinant clones homologous to the X chromosome. This consists in re-cloning

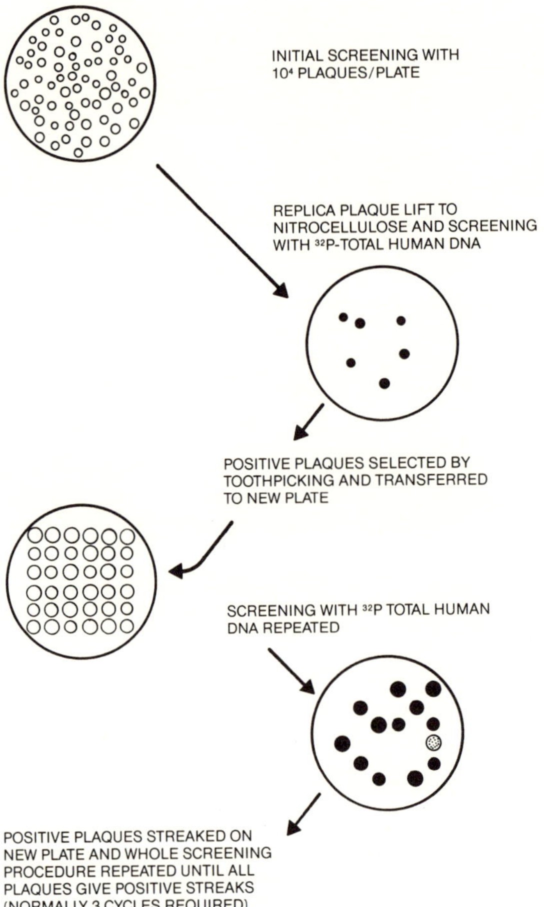

Figure 40: This figure summarizes the currently
adopted strategy of Gusella et al (44) with which
phage recombinant DNA clones including human repetitive DNA inserts can be easily separated from
those containing murine ones. The isolation of
the clones with human inserts is usually completed
by repeating the selection process for at least
three cycles. As it can be seen from Figure 41,
the clones isolated in this way usually contain
one repetitive human DNA sequence and several non-
repetitive ones which, given the structure of the
hybrid cell line used, are very likely to be unique
sequences of the human X chromosome. The sequences,
usually subcloned in a plasmid vector, are used
as probes for the screening of X-linked RFLPs.

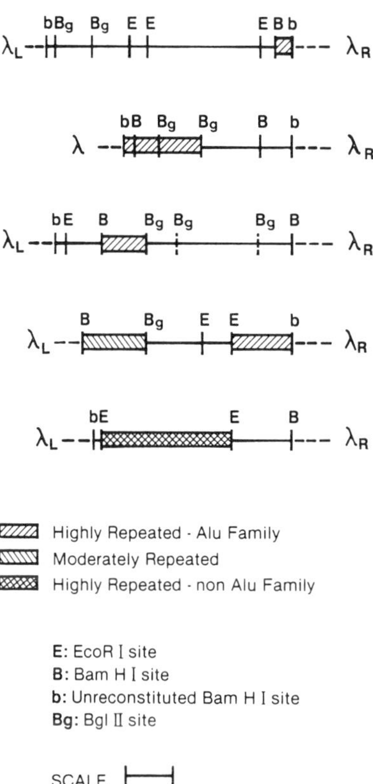

Figure 41: Characterization of the human X DNA inserts in five DNA recombinant clones from the library Charon 28: A9/HRBC2. The EcoRI – Bam HI 1.2 kb of the bottom clone has been subcloned in pBR322 and mapped to the region Xq13 → Xq24.

blindly the set of human X DNA clones isolated with the experimental strategy described above. After this second round of cloning, the isolation of the recombinant clones carrying single copy fragments of the human X chromosome can be achieved by screening for the subpopulation of recombinant clones which fail to hybridize with repetitive labeled human DNA as well as with phage DNA.

The strategies for the mapping of our probes to segments of the X chromosome has been discussed in detail elsewhere (108). Figure 40 outlines the hybrid cell panel approach and presents the panel of murine-human hybrid cells which we have available for mapping probes to 10 segments of the X chromosome. The assignment of single copy sequences by direct _in situ_ hybridization with diploid chromosomes has been successfully achieved by independent studies (109,110). Both approaches are routinely used in our laboratory. The former approach which gives a more precise localization will be most often used; although the latter approach used with prophase chromosomes prepared from cells containing X-autosomal translocations may be equally precise.

Subregional Mapping of the Single Copy Random X Chromosome Sequences

The hybrid cell DNA samples to be used in these panels are prepared from solid tumors grown in nude mice. Normally, subcutaneous injections of 10^7 hybrid cells will yield a 1-2 gram tumor in 30 days. DNA from each line is digested with EcoRI, separated on 0.8% agarose gels, denatured and transferred to nitrocellulose blot can be used five or more times without difficulty. Thus, one hundred probes could be mapped using as few as fifteen to twenty blots.

Mapping by the hybrid cell panel approach is straightforward and entails determining the smallest segment of the X chromosome which retains the restriction fragment(s) homologous to the probe. The probes used for these studies can be either single copy X specific fragments excised for gels or subcloned fragments. Since the X-autosomal translocations used to generate these hybrid cells come from different individuals, it is conceivable that some very common RFLPs may also be discovered during these subregional mapping studies. The primary goal, however, of subregional mapping is to identify which single copy sequences are near to each other in the genome. If two single copy sequences which identify RFLPs map to the same region of the X chromosome, it is likely that the RFLPs are within measurable linkage which will simplify the generation of a genetic map.

Using this approach, the EcoRI - Bam HI 1.2 kb unique sequence included in the recombinant phage clone at the bottom of Figure 41 has been mapped to the region Xq13 → Xq24 (S. Kahane, P. Szabo and M. Siniscalco, in preparation).

Preliminary data indicate that common multiallelic RFLPs exist among
Sardinian males for at least three restriction enzyme DNA sites
flanking the sequence or within it. Family studies are now in progress to establish the precise location of this DNA marker with respect
to the G6PD cluster (including Becker muscular dystrophy) and the
gene for Fabry's disease which is subregionally mapped to the interval
Xq22 → Xq23.

Currently, we have in storage white blood cells from the informative individuals of 22 three-generation pedigrees segregating
for Duchenne muscular dystrophy and
two for Becker muscular dystrophy as
Population Studies and   well as for other X-linked markers,
Linkage Analysis   notably Xg types and G6PD deficiency
(Table 11). Because this family material is relatively precious,
we do not intend to use it until a number of bonafide probes are
available. As soon as a number of probes have been shown to detect
useful RFLPs in the vicinity of either of the MD loci, we will screen
the progenitors of our Sardinian MD pedigrees for the presence of
restriction fragment length variants (RFLPs).

A unique feature of our Sardinian pedigrees is that they segregate for a series of X-linked genes whose location is often known
and covers the entire length of the X chromosome. It is for instance
well established that the loci for G6PD, deutan and protan colorblindness, hemophilia A, Lesch-Nyhan syndrome, Becker muscular
dystrophy and X-linked mental retardation, are all located between
Xq26 and Xqter (111). (Figure 42). The locus for X-linked ichthyosis
is located instead between Xp→ and Xpter (112) and is known to be
about 10 cM away (i.e. about 10^7 kb) from the locus of the Xg(a)
blood group (113). Because we have chosen to subregionally localize
our single copy probes for RFLPs, it is likely that we will be able
to select some within measurable linkage of known X-linked loci.
When enough RFLPs are mapped, we should also be able to tie together
the above mentioned three clusters of X-linked loci in a single,
comprehensive map which spans the entire human X chromosome and thus
be able to accurately map any X-linked locus including the muscular
dystrophy loci.

In collaboration with Drs. P. Soriano and G. Bernardi (IRBM,
Paris), Dr. P Szabo has undertaken a study to localize the multigene
family coding for actins in the human
and mouse genomes. Initial results
Identification of an X-   based on _in situ_ molecular hybridi-
Linked Actin Gene and its   zation studies indicate that the actin
Possible Value in Studies   genes are widely dispersed in both
on Becker Muscular Dystrophy these genomes. In the case of the
human genome, a total of 29 regions which contain actin genes or
pseudogenes were identified (114). One of these regions is located
in the distal third of the long arm of the X chromosome.

TABLE 11

List of Informative Sardinian Pedigrees
Segregating at Two or More X-Linked Loci*

Fifty-seven pedigrees segregating for G6PD deficiency, Xg-types
and/or deutan or protan colorblindness.

Eighteen pedigrees segregating for hemophilia A (HA), Xg-types and/
or G6PD deficiency or colorblindness.

Twenty-four pedigrees segregating for X-linked ichthyosis (ICT),
G6PD deficiency and for Xg types.

Thirty pedigrees segregating for Duchenne muscular dystrophy (DMD),
Xg types and/or G6PD deficiency or colorblindness.

Two pedigrees segregating for Becker muscular dystrophy (BMD) and
Xg types.

One pedigree segregating for ectodermal dysplasia anhidrotica (EDA),
G6PD deficiency and Xg types.

One pedigree segregating for hemophilia B (HB), G6PD deficiency and
Xg types.

Seventeen pedigrees segregating for X-linked mental retardation,
Xg types and/or G6PD deficiency.

One pedigree segregating for Lesch-Nyhan syndrome.

One pedigree segregating for Hunter's syndrome, G6PD deficiency
and Xg types.

*See reference 126 for further details on this family
material and its usefulness in the mapping of X-linked
molecular probes.

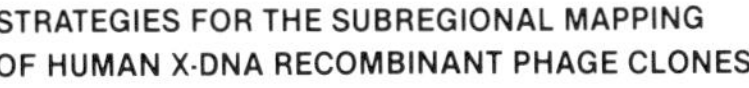

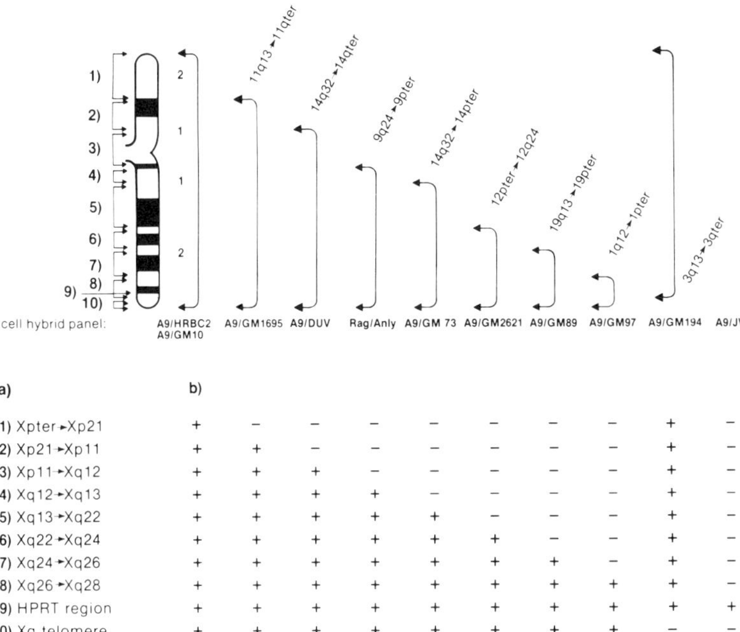

a)   b)

	A9/HRBC2 A9/GM10	A9/GM1695	A9/DUV	Rag/Anly	A9/GM 73	A9/GM2621	A9/GM89	A9/GM97	A9/GM194	A9/JW
1) Xpter→Xp21	+	−	−	−	−	−	−	−	+	−
2) Xp21→Xp11	+	+	−	−	−	−	−	−	+	−
3) Xp11→Xq12	+	+	+	−	−	−	−	−	+	−
4) Xq12→Xq13	+	+	+	+	−	−	−	−	+	−
5) Xq13→Xq22	+	+	+	+	+	−	−	−	+	−
6) Xq22→Xq24	+	+	+	+	+	+	−	−	+	−
7) Xq24→Xq26	+	+	+	+	+	+	+	−	+	−
8) Xq26→Xq28	+	+	+	+	+	+	+	+	+	−
9) HPRT region	+	+	+	+	+	+	+	+	+	+
10) Xq telomere	+	+	+	+	+	+	+	+	−	−

Figure 42: On the upper part of the figure the brackets on the left hand side of the chromosome sketch indicate the ten regions of the human X chromosome that are screened by the hybrid cell panel. The brackets on the right hand side describe the regions of the X chromosome exclusively retained by each of the hybrid cell lines listed. The notations next to these brackets (from A9/GM 1695 onwards) specify the autosomal fragment retained, in each case, along with the X-translocation chromosome which includes the human HPRT gene. The latter gene, presumably with its flanking regions, is the only portion of the human genome retained by hybrid A9/JW, as a result of chromosomal rearrangement in culture. The lower part of the figure summarizes: (a) the regions of the human X that can be screened with the use of the above hybrid cell panel; (b) the results expected when a DNA nonrepeated sequence homologous to a given human X DNA probe, is included within the indicated regions. (+) and (−) signs stand for "presence" or "absence" of molecular hybridization after exposure of the cell hybrid DNA to the labeled probe.

The unambiguous localization to the X chromosome was possible
because we used metaphase cells prepared from a human fibroblast
strain (GM 74) which contains two doses of a 14:X translocation
chromosome (14pter>14 q 32::Xq13>Xqter) which is particularly
easy to recognize because of its morphology even after _in situ_
hybridization studies. Figure 43 shows the distribution of silver
grain clusters (>3 silver grains/chromatid) along 25 of these
chromosomes cut from different metaphases. Although it is difficult
to localize a gene with precision when using ^{125}I labeled probes,
it is clear that the majority of the labeling is on the X chromosome
portion of the translocation chromosome. The most probable location
of the gene site is at Xq23→25.

Southern blots of EcoRI digested mouse, human and hybrid cell
DNA hybridized with the actin cDNA probe did not reveal any detectable extra human actin fragments in hybrid cells containing only
the human X chromosome. However, the detection of DNA polymorphism
with this probe is difficult because of the presence of the numerous
mouse actin bands. When Southern blots of EcoRI digested XY male
DNA and DNA from a XXXXX cell line are compared, only a 6.4 kb fragment shows the expected dosage dependence on the number of X chromosomes, thus confirming the X-linkage of one actin gene in humans
(Figure 44).

We are currently attempting to purify this human actin gene
from our hybrid cell libraries with the multistep procedure summarized
in Figure 45. From the gt7:A9/HRBC2 library, we have isolated twenty
actin positive clones. Sixteen of which are plaque purified. Of
these twenty clones fifteen label with total mouse DNA, three label
lightly with total human DNA and with mouse DNA, the remaining five
clones probably contain little or no highly repeated DNA sequences,
either mouse or human. We are now in the process of testing the
last five clones and the three which label with both probes to
determine if any of these contain the X-linked actin gene. The
purpose of this latter type of experiment is to isolate non-repeated
human DNA sequences that happen to be adjacent to the actin gene.
These unique sequences will in turn be used as DNA probes for the
screening of RFLPs in the same region which presumably also contains
the gene for Becker muscular dystrophy.

As it is well known to the members of this symposium, two
major forms of muscular dystrophy are inherited as X-linked recessive
traits in humans, Duchenne type
muscular dystrophy (DMD) and Becker
type (BMD). Two additional types,
Emery-Dreifuss-Hogan and Van Wyngaarden
are also X-linked. Linkage analysis

Current Knowledge of X-Linked Muscular Dystrophies and Future Prospects

in pedigrees segregating for BMD and G6PD or deutan colorblindness
or Xg blood groups have shown that the BMD locus is not genetically
linked to Xg, but is linked to the G6PD:deutan gene cluster with a

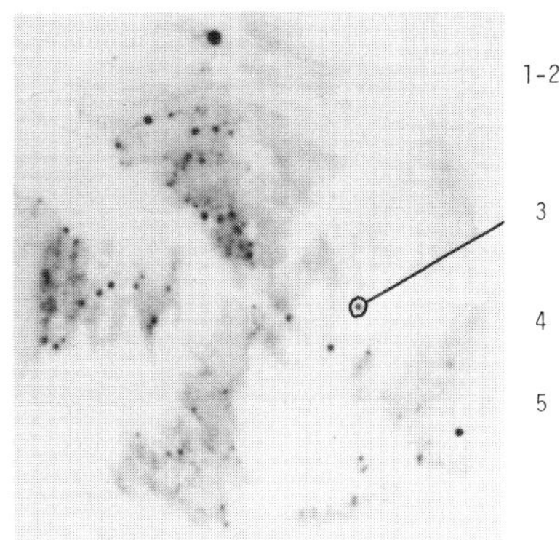

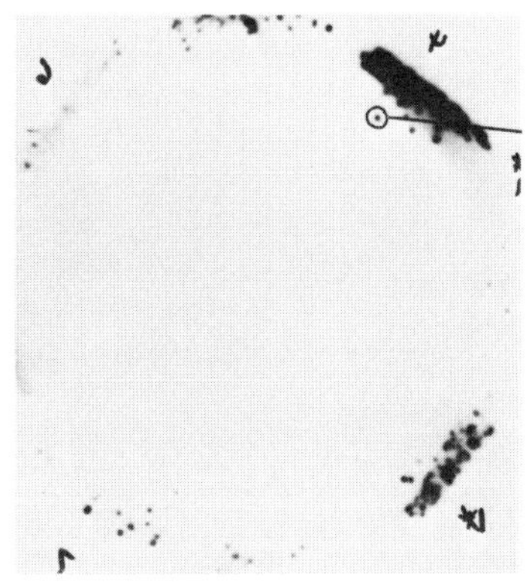

Figure 43: Screening of the X chromosome library
(gt7: A9/HRBC2) with cDNA plasmid clones:

1.   Library is transferred onto nitrocellulose.
2.   Southern transfer is screened with ^{32}P cDNA
     plasmid of choice.
3.   Positive plaques are picked up and re-streaked
     on new bacterial lawns.
4.   Steps 2 and 3 are repeated twice.
5.   Positive plaques are re-picked and re-screened
     for their homology with the labeled probe.
6.   When streaks are uniformly positive, isolated
     plaques are re-picked for preparation of plate-
     lysate and growth of pure phage stock.

This strategy was applied for the isolation of phage
recombinant clones (from our library gt7: A9/HRBC2)
containing unique DNA sequences homologus to the G6PD
cDNA plasmid of Persico et al (119) and the cDNA
plasmid of Soriano et al (114).

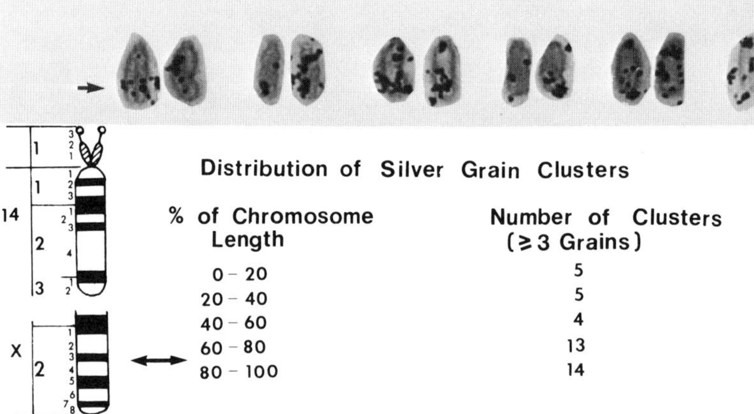

% of Chromosome Length	Number of Clusters (≥ 3 Grains)
0 – 20	5
20 – 40	5
40 – 60	4
60 – 80	13
80 – 100	14

Figure 44: Localization of the X-linked actin gene.
^{125}I labeled Syrian hamster cytoplasmic cloned actin
cDNA was annealed to metaphase chromosomes from a
cell strain bearing an X:14 translocation. Several
examples of the X:14 translocation chromosome are
shown with clusters of silver grains seen near the
telomere of the translocated X chromosome long arm.
Often both chromatids are labeled at homologous
positions. The distribution of silver grain clusters
(≥ 3 grains) over 25 translocation chromosomes is
reported. The region Xq23→q25 is the most probable
location of the actin gene.

maximum likelihood estimate of the recombination fraction of 0.27
for the distance from BMD to G6PD (115). These results strongly
suggest that BMD is within the subterminal region of the long arm
of the X chromosome, since G6PD has been localized to the terminal
Giemsa negative band Xq28 (116). DMD does not appear to be linked
to either the G6PD: deutan cluster or the Xg locus; thus, DMD and
BMD are separate loci. Five sporadic cases of females affected by
DMD have been reported to be associated with an X chromosomal rearrangement involving the short arm region Xp21 (95,96,117,118). These
findings suggest that the structural gene for DMD may be located
at that site.

From what I summarized above and from the individual presentations of Drs. Bruns, Caskey, Gusella, Housman, Kunkel, Latt, White
and Williamson at this symposium, it is clear that in the next few
years many single copy, human chromosome specific probes will be
isolated, especially for the X chromosome. We anticipate that our
group alone will isolate 30-40 such X chromosomal probes this year,
many of which can be used to detect RFLPs. Some of these RFLPs are
expected to be near the DMD loci and the BMD loci and should prove
to be invaluable tools for mapping the MD loci.

One further factor to consider is that the recombination
frequency near ends of both arms of the X chromosome may be higher
than expected on the average recombination frequency genome. This
means that one requires more markers to make an accurate linkage
map, but also that if one has a closely linked marker to a deleterious
gene, it may be possible to "walk" the genome and ultimately isolate
the gene responsible for the defect.

The genes coding for G6PD and HPRT have been reasonably well
localized cytologically (G6PD at Xq28, HPRT at Xq26-27) and are not
separated by more than ten percent of the length of the X chromosome
or approximately 1.5×10^7 bp as an estimate. Yet these two markers
are genetically unlinked, that is separated by >50 cM. In this
region the maximum number of nucleotides per cM would then be
3×10^5 bp. If one of the RFLPs identified by our randomly selected
single copy probes were located within 1-2 cM of the BMD locus, we
could "walk" through the BMD gene using a cosmid DNA library. The
same would also be true for the DMD locus.

For mapping BMD it is clear that the cDNA clones of its
neighboring genes such as the cDNA clone for HPRT, announced at
this meeting by Dr. Caskey and that for human G6PD reported by
Persico et al (119) may turn out to be of particular value to
screen for X-linked RFLPs within measurable genetic distance of
the BMD locus.

We have isolated genomic clones containing the gene coding for
G6PD from our gt7:A9/HRBC2 recombinant DNA library. This was done
using cloned cDNA probes, p6405 and
p6222, (provided by Dr. Luzzato) (119)
which have been shown to contain cDNA
for G6PD mRNA by positive translation

Isolation of Human and
Murine Genomic G6PD Clones

assays. The purification protocol is outlined in Figure 45. One
of 10^5 plaques, seven positive plaques were isolated. Of these
five remained positive in the second screening. Four of these five
(G1, G4, G5 and G7) have now been plaque purified; purification
of the last one (G6) is in progress.

The four purified phages were of three types; G1 and G7 were
identical and contained a single 6.7 kb EcoRI insert fragment.
G4 contained a 6 kb and a 3 kb EcoRI fragment and G5 contained only
the 6 kb fragment. By hybridization of EcoRI digests of these phages
with total human DNA; the 6 kb fragment contains only low or single
copy sequences. This result suggests that G4 contained the human
G6PD gene and that G5 which also contains the 6 kb fragment was
probably of human origin. G1 and G7 contain the mouse G6PD gene
within 7 kb. A preliminary screening done with the G5 probe suggests
that there are several EcoRI RFLPs found at appreciable frequencies.

Given the already mentioned linkage between the G6PD cluster and
BMD, it has to be expected that RFLPs identified through the non-
repeated DNA sequences isolated from the above recombinant phages,
should also be linked to the BMD locus and could therefore be useful
probes for mapping studies as well as for "walking" studies to isolate
the presumed structural gene for BMD.

Studies on the molecular mapping of the human sex chromosomes
X and Y are particularly rewarding in view of the circumstance that
both of them occur in haploid condition
in the male sex, thus facilitating the
screening for RFLPs and their sub-

Ultimate Goals

regional mapping with the use of 3-generation pedigrees with instances
of recombination, chromosomal rearrangements and/or non-disjunction.
This general issue is the goal of a much broader research endeavor
of our group. The finding of X-linked RFLPs in linkage disequilibrium
with one or the other type of MD will obviously offer an indirect
way for the fine molecular mapping of the structural genes responsible
for these diseases without the need of isolating them. From a medical
standpoint such a finding offers a very powerful tool for the detection
of silent heterozygous carriers and for the prenatal diagnosis of
the disease in fetuses. Furthermore, when population studies will have
established the type of DNA variants regularly associated (as a result
of linkage disequilibrium) with each MD mutant, it will be possible
to use such criteria to prove the occurrence of de novo mutations in
families with sporadic cases and to approach -- at a fine molecular
level -- the problem of genetic heterogeneity between unrelated
patients from the same or different populations.

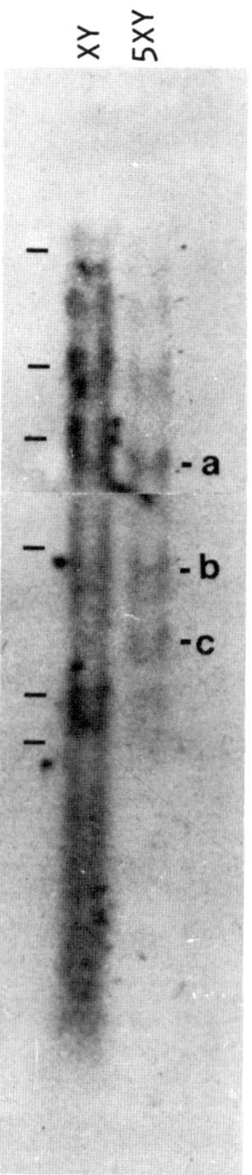

Figure 45: X chromosome dosage dependence of an
actin gene containing EcoRI fragment. DNA from a
normal male (XY) and a fibroblast cell line with
five X chromosomes were digested with EcoRI and
separated on a 0.6 percent agarose gel. The
Southern transfer of this gel was hybridized with
a Syrian hamster cytoplasmic actin cloned cDNA
probe. There was approximately three times as much
DNA in the XY lane so that the autosomal actin bands
are less intense in the XXXXXY lane. Fragment a is
labeled to about the same extent in both lanes, thus
suggesting a dosage effect and therefore X-linkage
of this actin sequence. Fragments b and c are seen
with all plasmid probes and are due to the hybridization of pBR322 DNA to mitochondrial DNA which was
present at a higher level in the XXXXXY DNA. The
dashes indicate the positions of the λ Hind III
fragments: 23.5, 9.7, 6.6, 4.3, 2.2 and 2.1 kb.

DR. LARRY SHAPIRO:* I would like to share with you some
of our recent observations on the regulation of X chromosome inactivation. While this research is
not directly related to the problem
of muscular dystrophy, I hope it will
still be of interest to this group.
Perhaps our findings will eventually
have some practical relevance to considerations of carrier detection
in Duchenne dystrophy by increasing our understanding of the
mechanism of X chromosome inactivation. Once the muscular dystrophy
gene product is identified, it is conceivable that the ability to
regulate the expression of an inactive X chromosome in vitro would
be an adjunct to various strategies for heterozygote identification.

X Chromosome Inactivation
and Reactivation by DNA
Methylation Probes

During the past two years, we have tried to develop a model in
a human-rodent hybrid cell system in which we can isolate a structurally normal but functionally inactive X chromosome following
segregation of a cytogenetically and biochemically marked active
X from the somatic cell hybrid. In this manner we can study the
properties of the inactive X and perturb it in a variety of controlled
ways.

The general strategy has been to produce hybrids between hamster or mouse HPRT⁻ established cell lines and normal diploid human
cells from females carrying balanced X-autosome translocations. As
we have heard several times during this meeting, it is usually the
structurally normal X chromosome which is observed to be inactivated
in such situations. In the particular example I will focus upon,
we have made use of a fibroblast cell line from a female carrying
an X-11 translocation (121). The breakpoints in this reciprocal
translocation are shown in Figure 46. There is a structurally
normal X, a normal 11, and two rearranged chromosomes. It is of
some interest that the donor of this cell line had a clinical diagnosis of Duchenne muscular dystrophy consistent with the location of
the breakpoint in the X chromosome at Xp21 as we have heard from
Dr. Francke. A variety of hybrids were isolated following fusion
and selection in HAT medium and were scored for chromosome constitution and X chromosome markers. As anticipated, all of the hybrids
retained the X-11 chromosome as this was the earlier replicating X
in the patient and presumably provided the active HPRT gene. One
very interesting hybrid designated 37-26 was observed to have both
the X-11 chromosome and the normal but inactive X. This line was
positive for a number of human X-linked markers including STS. This
is consistent with our previous assignment of this locus to the
distal X chromosome short arm (122) and the continued expression of
STS from an otherwise inactive X chromosome (123).

*Work conducted in collaboration with T. Mohandas at Harbor-UCLA,
 and also with Robert Sparkes, UCLA and Peter Jones and Shirley Taylor
 U. of Southern California and Lee Venolia and Stanley Gartler at the
 University of Washington.

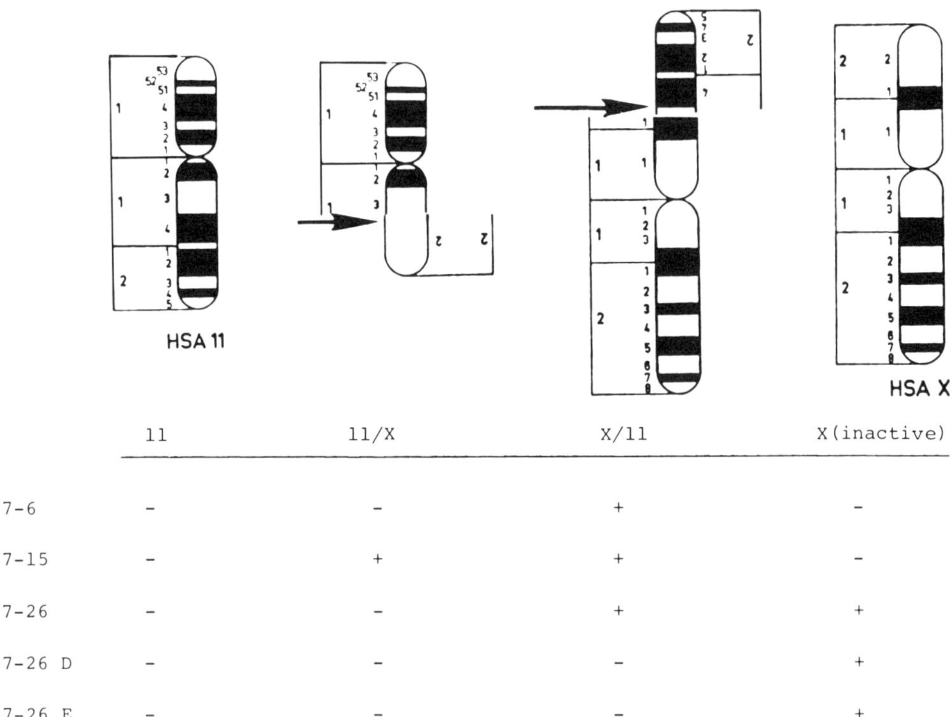

	11	11/X	X/11	X(inactive)
37-6	−	−	+	−
37-15	−	+	+	−
37-26	−	−	+	+
37-26 D	−	−	−	+
37-26 E	−	−	−	+

Figure 46: Schematic representation of the breakpoints in the X-11 translocation cell line (GM 1695)
and the chromosome constitution of several hybrid
clones produced by fusion of GM 1695 and mouse A9
cells. The fibroblast line derived from a female
with a 46,X,t(X;11) (Xqter>Xp21::11q13>11qter;
11pter>11Q13::Xp21>Xpter) karyotype, was fused
with an HPRT⁻ mouse cell line (A9) and growth of
hybrids was selected for in HAT. A series of
clones was isolated and characterized cytogenetically and biochemically. 37-26D and 37-26E were
derived by back selection of 37-26 in medium
containing 8-azaguanine.

This must have been the case since 37-26 did not have the reciprocal
translocation product (11-X) which we had shown to carry the other
STS allele in this patient. The hybrid line 37-26 was then placed
in medium containing 8-azaguanine to select for cells which had lost
the active copy of the HPRT gene, and a series of subclones was
derived which in fact had lost the X-11 but retained the cytologically
normal X. As anticipated, expression of a number of standard human
X chromosome markers including HPRT, G6PD and PGK were not detected
in these hybrid subclones, however, human STS continued to be produced. Using the BrdU-acridine orange staining protocol to study
the timing of replication of the human X in these hybrids, Dr. Mohanda
was able to show that even in this situation in which most of the othe
human chromosomes had been segregated, the X retained its late pattern
of DNA synthesis and BrdU incorporation. Thus, we had available a
cell line with a structurally normal but inactive human X chromosome
isolated from active X material and from most of the human autosomes.
Similar cell lines have been produced in other laboratories (124,125)
and we have identified several others as well with a variety of human
X chromosomes in either a mouse or a Chinese hamster background. Such
hybrids have permitted us to investigate the mechanism of X inactivation by perturbing the inactive X in a variety of ways to see if
we could induce reexpression of human X-linked genes.

One of the models that has been proposed to account for many of
the observations regarding X inactivation involves the regulation
of expression of X encoded genes by DNA methylation. This idea
was proposed by Art Riggs (126) at the same time that Holliday and
Pugh (127) were considering the potential role of methylation for
control of developmentally regulated genes. It is particularly
attractive to consider methylation as a potential mechanism involved
in X inactivation because of a number of biological observations
attesting to the considerable stability of X inactivation patterns.
Once established as an inactive X chromosome, an X is almost never
observed to undergo spontaneous reactivation even following many
cell divisions. One might intuitively assume that if X inactivation
were maintained by DNA-protein interactions, these associations would
be transiently disrupted by DNA replication and result in loss of
clonal stability of X inactivation.

The initial experiments in our system to examine the role of
methylation involved the use of 5-azacytidine as originally described
by Jones and co-workers (128,129). When 37-26 cells are grown in
HAT selective medium to look for HPRT expression, we observed HPRT$^+$
cells arising with a frequency of 10^{-6} to 10^{-7}. This usually
corresponds to spontaneous reactivation of the human HPRT gene as
determined by isoelectric focusing, but rare expression of mouse
HPRT, presumably corresponding to reversion at the mouse locus was
also observed. Thus, when about 100,000 cells are plated per
dish and the cells switched to selective medium, colonies rarely
appear. However, if the cells are treated with 5azaC for 24 hours,

allowed to recover for 72 hours and then placed in HAT medium, many
HPRT positive colonies appear (130). For this particular cell line,
this effect occurs at an optimal dosage of 5azaC of 2-3 µM (Figure
47). The data shown in Figure 47 have not been normalized for
cytotoxicity of 5 azacytidine as have some of the later data to
be provided. The variable cytotoxicity between cell lines needs
to be taken into account when evaluating 5azaC effects and so under
optimal conditions perhaps 1 in 100 cells surviving treatment is
reactivated at the HPRT locus.

We have examined the ability of a number of other compounds
and treatment protocols to similarly reactivate HPRT. 6 azacytidine
bromodeoxyuridine, cytosine arabanoside, and several other drugs
are ineffective. However, in collaboration with Peter Jones and
Shirley Taylor, we have found 5 azadeoxycytidine to be effective
at 1/10 the concentration of the riboside. Pseudoisocytidine is
also a potent inducer of HPRT expression. In fact, with the latter
compound, we have been able to get eight percent of surviving
cells to express HPRT.

In all of these experiments, we have randomly picked and
examined a considerable number of "reactivants" to document by
isoelectric focusing that the HPRT expressed is in fact of human
mobility. This has uniformly been the case. By picking colonies
and expanding them to suitable numbers of cells, we have also been
able to use a variety of enzyme electrophoretic systems to look
for concommitant reactivation of the other human X encoded loci.
Specificially, we have investigated G6PD, PGK and α-galactosidase
A expression. In the latter instance, no suitable electrophoretic
procedure exists for separating mouse and human isozymes, but we
have been able to make use of a human specific α-gal A antibody
as described by Bishop et al (131) with an antibody kindly provided
by Dr. Robert Desnick. We have found either G6PD or PGK to be
reactivated in five to seven percent of colonies induced by
5azaC to express HPRT. α-gal A expression occurs at a considerably
higher frequency. However, it would appear statistically that
reactivation of these loci occurs essentially independently. From
these results we infer that at least the maintenance of X inactivation
occurs in a segmental fashion although this clearly does not pre-
clude a chromosomal event, perhaps involving a single inactivation
center, at the initiation of the inactivation process.

Another aspect of the reactivation of X-linked genes by 5azaC
is the stability of these events. Exposure to 5azaC in our experiments is for one cell doubling time or less, and we have observed
the phenotype of at least HPRT$^+$ and G6PD$^+$ to be stable in the
absence of 5azaC through many cell divisions with or without
continuous selective pressure for HPRT (130).

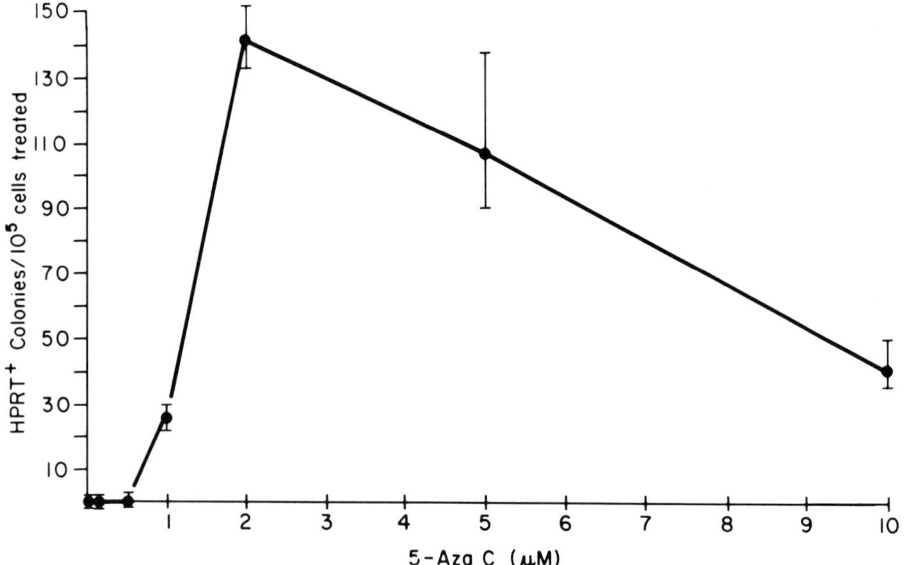

Figure 47:   Dose response of HPRT induction vs.
5-azacytidine concentration.   100,000 37-26D
cells were plated into dishes, exposed for 24
hours to the indicated concentration of 5-aza-
cytidine, and allowed to recover for 72 hours
in non-selective medium.   They were then placed
in HAT and surviving colonies were counted two
weeks later.   These data are not corrected for
5-azacytidine cytotoxicity.

Recently, we have done another series of experiments involving the use of DNA mediated transformation of the HPRT gene (132). We have asked whether DNA from hybrid cell lines containing an active X chromosome, an inactive X chromosome and a reactivated X chromosome function equally well in such an assay system. We have utilized two different HPRT⁻ recipient cell lines. When we used DNA from a hybrid cell containing an active X chromosome, we got the expected frequency of transformation of HPRT (Table 12). However, DNA from the hybrid cell line containing the inactive X, did not confer HPRT⁺ on the recipient cell line. When the HPRT locus was reactived by 5 azacytidine treatment, however, the DNA regained its ability to transfer HPRT. As a control, we examined TK transformation to ensure that all of the DNA preparations were of equal quality (Table 12). All of the DNAs including the prep from the cell line with the inactive X, were equally effective in TK transfer to deficient recipient cell lines. These results confirm that some event at the DNA level, is responsible for X inactivation.

In other studies (133) we have examined the cell cycle specificity of the 5 azacytidine effect. We synchronized the cell line with the inactive human X chromosome by a mitotic shakeoff procedure. The degree of synchrony was verified by ^{3}H thymidine incorporation studies. We then pulsed the cells for two hour intervals with 5 azadeoxycytidine at various times during the cell cycle. We observed the induction of HPRT⁺ "reactivants" only when the drug was delivered during the S phase, and the maximal effect occurred during the latter half of the S phase.

We next asked whether the length of the recovery period between 5azaC treatment and exposure to HAT selective medium had any bearing on the frequency of HPRT⁺ colonies. Our observation was that if cells were allowed tc go through two doublings prior to selection a maximal frequency of reactivation was found. This fits very well with the proposed model for the effects of 5azaC.

Methylated cytosine constitutes three to seven percent of the cytosine residues in mammalian DNA and ninety percent of these modified bases are found 5' to a G (134). There is also almost always symmetrical methylation around $\frac{CG}{GC}$ sites. This is due to the action of so called maintenance methylases which have a marked preference for hemimethylated DNA and so preserve a pattern of methylation throughout many rounds of DNA synthesis, once that pattern has been established. Azacytidine treatment followed by incorporation into DNA probably inhibits methylase activity and results in random demethylation which becomes fixed after two cycles of replication.

In summary, we feel we have a system of some interest for looking at the mechanism of X inactivation. This is possible because of the hybrid system's ability to segregate the active

TABLE 12

HPRT TRANSFORMATION

Recipient A

	Frequency
No DNA	0
Hybrid with active X	2.5×10^{-6}
Hybrid with inactive X	0
Hybrid with reactivated X	0.5×10^{-6}
Hybrid with reactivated X	0.6×10^{-6}

Recipient B

No DNA	0
Hybrid with active X	$.12 \times 10^{-6}$
Hybrid with inactive X	0
Hybrid with reactivated X	$.13 \times 10^{-6}$

Note: These experiments are described in reference
153. High molecular weight DNA was isolated from the
donor cell lines indicated by standard methods and
applied to cells by calcium phosphate co-precipitation
technique.

from the inactive X chromosome. My personal bias is that DNA
methylation does not completely explain X inactivation. Methylation
is clearly a process used to control gene expression around the
genome and so is too general to be the sole explanation for X inactivation. Rather, methylation may serve as a segmental, "locking
in" mechanism for inactivation initiated by some as yet poorly
understood chromosomal process. Hopefully, as some of the cloned
X chromosome probes being described at this meeting become available
for study, the expression of these segments can be monitored and
direct molecular verification of the role of DNA methylation in X
chromosome inactivation can be achieved.

DR. SCHWARTZ: You said methylation controls expression. That
has not been proven yet. Moreover, in Drosophila, which has a
very low degree of methylation, there is no direct correlation with
methylation and gene expression.

DR. SHAPIRO: You are quite right. Drosophila appear to have
little or no methylcytosine, but regulation may clearly be different
in lower eukaryotes. I agree that a direct causal relationship
cannot yet be said to exist, although our data suggest that by
interfering with methylation in this system we can, in fact,
influence expression.

DR. KUNKEL: Have you looked at the replication pattern of
that inactive X that is reactivated and can you see little areas
lighting up?

DR. SHAPIRO: We think we can. They are not consistent from
clone to clone and again one might not expect them to be. But
there are minor alterations that are suggestive. We haven't got
the slightest idea how long a segment of DNA we are interfering
with here and whether one could even expect to see that. You are
raising a very interesting point which everybody in this room
probably accepts as a valid assumption, and that is the direct
and absolute correlation of late replication and lack of expression.

DR. LATT: About your hybrid. What cell did you use as the
parent with the X-11, was it a fibroblast or lymphocyte?

DR. SHAPIRO: A fibroblast line.

DR. LATT: Because of the late replication pattern for
reasons yet to be explained, the late X has a different pattern
of replication in fibroblasts than in lymphocytes, at least most
T cell lymphocytes. Hunt Willard was the first to publish on
this as a function of cell type and we have confirmatory data.

DR. SHAPIRO: How about if you take X from lymphoblast and
put it into a hybrid milieu?

DR. LATT: I wondered about that for a long time having been
technically unable to get the information. That is why I was
asking you if you did the same thing using the white cells from
the patient instead of fibroblasts. In the fibroblasts the late
replication highlights the distal part of the long arm. In most
of the lymphocytes, it does not.

DR. NADAL-GINARD: Have you tried to activate the X in a
heterozygote cell line, not in a hybrid? Let us say one with
G6PD deficiency?

DR. SHAPIRO: We have tried, and I know at least one other
laboratory that has tried as well in fibroblasts from a female
doubly heterozygous for G6PD and HPRT deficiency and neither group
was successful. I think this more likely than not was for
technical reasons. Most of the cell lines are diploid fibroblast
lines that have all been through one cloning and so they are kind
of old, tired cells. The other thing I alluded to a few minutes
ago is that azacytidine cytotoxicity is very cell type dependent.
In doing the reactivation experiments, you are dealing with essentially the therapeutic index of this drug. I have got a feeling
that perhaps in those fibroblast cell lines the therapeutic index
is working against us.

DR. NADAL-GINARD: It seems that your activated phenotype
is very stable. Have you tried to go through many rounds of azacytidine treatment to see how much of the X you can activate?

DR. SHAPIRO: We have got those experiments in progress right
now.

DR. SINISCALCO: Did I understand that you were questioning
the overlapping of late replication and expression. I thought
there was an interesting paper five or six years ago by Rattazi
and Cohen working on mules and hinnies and cloning their fibroblasts.
As you know, G6PD can be distinguished between horse and donkey and
they showed that whenever there was a late replicating horse chromosome, the donkey G6PD was expressed and the other way around.

DR. SHAPIRO: This is at a chromosomal level. The late
replicating X is not uniformly late replicating and I am not sure
one can correlate this necessarily with individual gene expression.
I think that would be another very interesting thing to do when
appropriate probes are available.

DR. LATT: This is something I suspect others in addition to
ourselves are very interested in. Dr. Kunkel is obtaining probes
on the X, some of which will probably be expressed, others will
not. It should be possible to test for the correlation between
replication and expression at the level of specific probes. The
best people have been able to do so far is to look at chromosomes,
using various BudR dye techniques to get correlations. Christine
Disteche has obtained such data in murine cells.

DR. DAVIDSON: Have you tried a variety of mutagens on these
cells to see if there are other mutations that will cause X reactivation?

DR. SHAPIRO: Yes, and certainly others more extensively than
we have. I am not aware of any of the garden variety mutagens that
have this sort of effect. However, we get into a little bit of
a semantic problem if a mutation is a stable alteration in DNA,
then alterations of methylation are a type or a class of mutations.

DR. LATT: I have been intrigued by one other factor which may
or may not be relevant to this which is that 5 azacytidine is known
to induce sister chromatid exchanges. There is a paper on this by
Banerjee, A., et al, CANCER RESEARCH, Vol. 39, p. 797-799, 1979.
I have wondered whether this says that methylation has something
to do with sister chromatid exchanges or perhaps that DNA inter-
change has something to do with what you are seeing. It is just
something to keep in the back of your mind, that you are already
using something that may have the properties that Dr. Davidson
mentioned.

DR. C. THOMAS CASKEY:* Our laboratory has approached cloning
X chromosome DNA sequences from two strategies. The first was
obtaining recombinant DNA sequences
Cloning the HPRT Gene   to specific gene, hypoxanthine-
guanine phosphoribosyltransferase
(HPRT). The second approach is similar to the approach that other
laboratories have reported here; cloning anonymous S unique sequences from special λ libraries. We have succeeded in obtaining
HPRT clones and have had initial success with X-linked anonymous
sequences isolation.

The HPRT locus in man and mouse and hamster is X-linked. On
the basis of human genetic studies, it is known to be activated
by lyonization. Because of the highly mutable character of this
locus in cultured cells and the spectrum of the clinically and
biochemically different Lesch-Nyhan cases, we undertook the study
of HPRT toward the objective of understanding the molecular basis

*In collaboration with Dr. A. Craig Chinault and Dr. Robert Nussbaum.

of mutation in both cultured cells and man. I was encouraged in
hearing Dr. Francke's most recent updating that the Lesch-Nyhan
cases which now indicates twenty percent new mutation frequency
among isolated cases. We recently obtained cDNA clones of the
mouse HPRT and found that these clones have DNA homology with HPRT
sequences of man and hamster. Thus, our murine cDNA clones permit
studies in man.

The _in vitro_ translation of Chinese hamster brain messenger
RNA is given in Figure 48. The total translation (Figure 48) is
given in lane A. A control lane (B) indicating the HPRT obtained
from cultured Chinese hamsters cells labeled with (S^{35}) methionine
and isolated by immunoprecipitation. In lane C, non-immune sera
was used. Lanes D and F had immunoprecipitation carried out with
antibody to native and denatured HPRT respectively. Thus, the
reticulocyte extract has the capacity to translate HPRT message
in vitro and permit native assembly. Initially, our source of
mRNA was hamster brain since it was known brain had the highest
level of message among body tissues. Later in our cloning efforts,
a mouse neuroblastoma cell line (NBR4) isolated by David Melton
was shown to overproduce HPRT mRNA.

The NBR4 line had many interesting properties. As shown in
Figure 49, the HPRT^{+} NB^{+} HAS A 12:X translocation. The HPRT^{-} clone
used to derive NBR4 had similar properties. However, certain HPRT^{+}
revertants (NBR4) had unusual cytogenetic and biochemical properties.
Figure 49 points out the most striking one.

As you can see NBR4 has an unusual chromosome with a single
centrimeric area, a 12:X translocation and a triplication of the X
leading to a large chromosome. This cell (NBR4) line was found to
overproduce HPRT _in vivo_ as determined by HPRT radioimmunoassay.

There are several points to be made from Table 13 with reference
to obtaining cDNA cloning. The wild type neuroblast (NB^{+}), (NB^{-})
cell line, and the HPRT^{+} revertant cell line (NBR4) are compared
for endogenous HPRT activity, CRM (cross-reacting material), and
mRNA. NB^{-} cells had no detectable HPRT activity, or CRM as determined by _in vivo_ measurement. However, the mRNA obtained from NB^{-}
produced translatable HPRT activity equal to the wild type cell line.
In the revertant NBR4 cell line, while enzymatic HPRT specific activity
was low, the relative amount of _in vivo_ detected CRM and _in vitro_
detected mRNA is quite high. Since the level was at least 10-fold
higher for the NBR4 over the level detected in hamster brain, we
elected to carry out cDNA cloning on NBR4 mRNA.

The cDNA cloning utilized pBR322 by G-C, tailing and insertion
into the plasmid Pst site. We took advantage of relative differences
of HPRT mRNA for NBR^{-} and NBR4 to identify HPRT recombinants. As

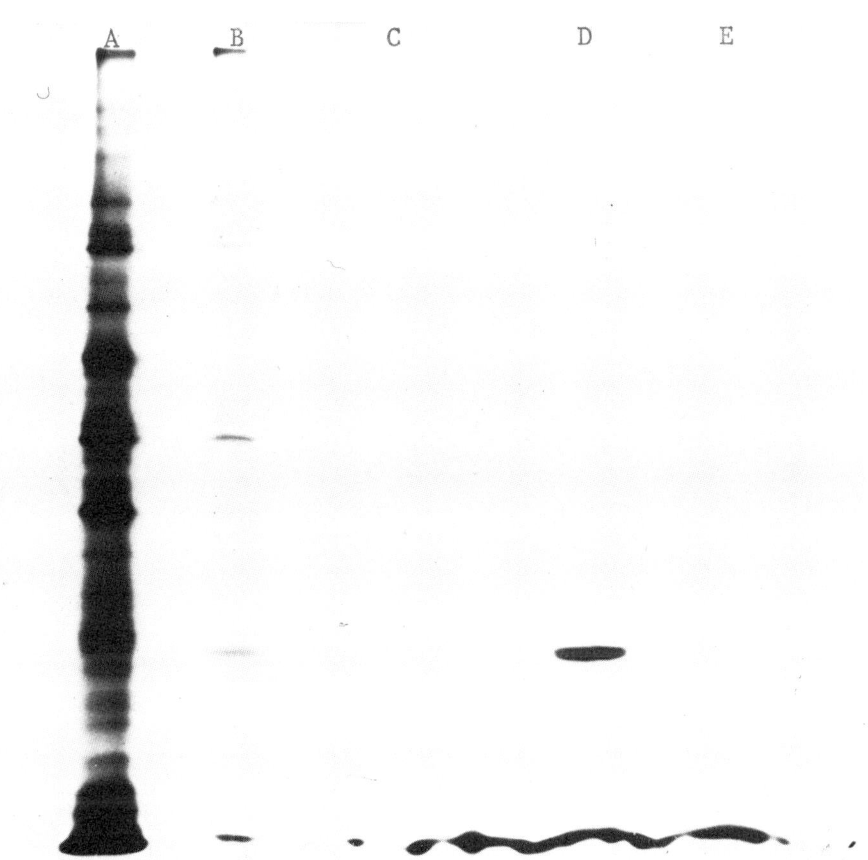

Figure 48: In vitro reticulocyte translation of
HPRT mRNA. Lane A) total hamster brain mRNA in
vitro translation products; B) immunoprecipitated
in vivo labeled hamster HPRT; C) non-immune sera
immunoprecipitated of in vitro hamster brain
mRNA in vitro translation products; D) native anti-
HPRT immunoprecipitated hamster brain mRNA in
vitro translation products. Analysis is by auto-
radiography of a SDS PAGE.

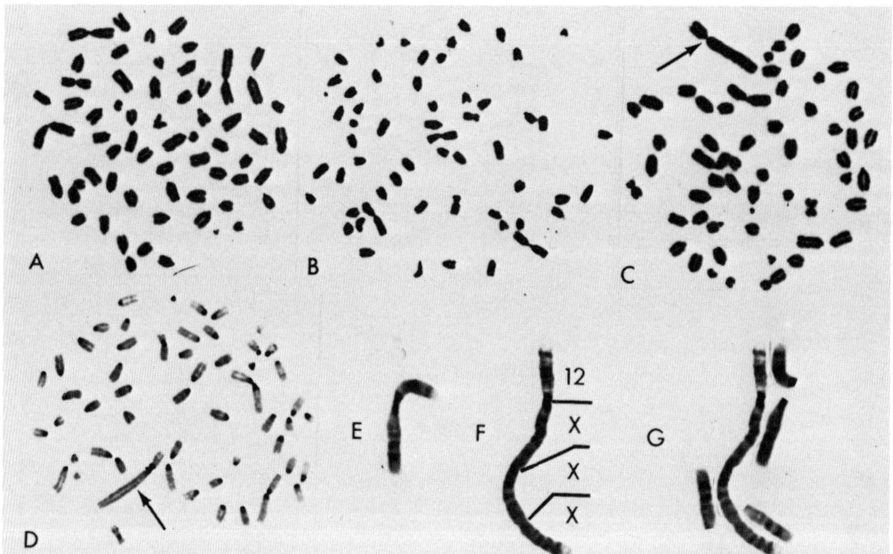

Figure 49:   Giesma-stained metaphase of (A) NB$^+$, (B) NB$^-$, and (C) NBR4.   (D) C-banded metaphase of NBR4.   (E) G-banded marker 3 from NB$^+$, comprising a centric fusion of chromosome 12 and the X.   (F) G-banded elongated marker of NBR4, comprising chromosome 12 and three copies of the X with the middle segment inverted.   (G) Reconstruction of the NBR4 marker shown in F from the marker 3 from NB$^+$ shown in E.

Table 13

Cell or tissue source	HPRT enzyme activity*	HPRT immunoprecipitable radioactivity	
		In vivo	In vitro
RJK0	1.55	1.80	1.35
RJK3	0.32	ND	1.12
RJK39	< 0.01	ND	0.89
RJK36	< 0.01	undetectable	undetectable
NB+	**0.83**	**0.98**	**0.96**
NB−	< 0.01	undetectable	0.67
NBR4	**0.74**	**9.33**	**47.40**
Brain	**6.80**	ND	**4.08**
Liver	0.03	ND	0.61
Testis	0.04	ND	0.67

HPRT enzymatic activity was determined from cell
extracts. HPRT *in vivo* CRM was determed by immunoprecipitation of *in vivo* labeled cells. *In vitro*
CRM was determined by immunoprecipitation of *in
vitro* reticulocyte extract synthesis of HPRT using
mRNA isolated from the respective cells.

shown in Figure 50, a large number of plasmid recombinants are
analyzed by their differential hybridization to the cDNA probes
from NB⁻ and NBR4.

We selected some 32 recombinants with features which were
not quite as positive as the recombinant indicated by the arrow.
Any recombinant which exhibited differential hybridization was
included. We grouped these plasmids and immobilized them to
millipore filters for HPRT mRNA selection.

Figure 51 indicates the results of our mRNA selection. As
you can see there was a weak signal detected when four recombinants
(pHPT1) immobilized to a single filter were used to select the
HPRT message. We then constructed four individual filters each
containing a single recombinant and were able to identify one of
the four recombinants as corresponded to HPRT.

We then examined the relative abundance of the message for
each of the four recombinants on the initial filter. In Figure
52, by Northern analysis, NBR4 did have an elevated level of HPRT
message while the level of mRNA for the NB⁻ and the NB⁺ appeared
equal. No differences were observed with the three additional
recombinants. Thus, by two methods we assigned HPT1 as an HPRT
cDNA recombinant. Southern analysis uses the same four recombinants.
The HPRT gene copy number is increased on the NBR4. It is interesting
that recombinant A, which codes for different gene is also amplified.
We have done additional studies with recombinant A. Since we had
triplication of the X in NBR4, it is possible that recombinant A
corresponds to a less amplified X sequence (Figure 53).

The endonuclease restriction map of HPT1 and a subsequently
identified recombinant HPR2 are shown in Figure 54. These recombinants occurred at a frequency of 1 in 1000. The HPT2 clone
represents an insert which is approximately seventy percent of the
estimated overall length of the mRNA for mouse. The 3' terminus
of the mRNA is to the right based on our initial DNA sequence work.
Thus, we have an incomplete cDNA clone for mouse at this time.

The Northern and Southern analyses of mouse, hamster and
human materials are given in Figure 55. It is clear from the
Northern analysis that NBR4 is over-producing mRNA. Furthermore
the high molecular weight RNA is not the template for HPRT. The
HPT2 does cross-react with baboon and hamster mRNA. On the right
side, the Southern analysis of the wild type mouse cell line,
the NBR4 over-producer, human and hamster DNA are given. The HPT2
probe clearly cross-reacts with three different species; we have
taken advantage of that in trying to understand the composition
of the gene and the understanding of mutants of this locus.

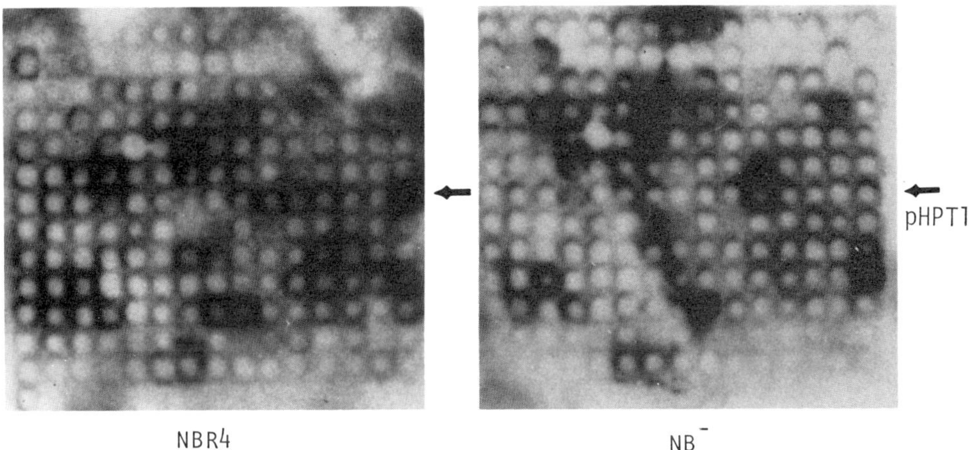

Figure 50: Differential hybridization of plasmid DNAs to (^{35}P) cDNA probes synthesized from NBR4 (HPRT overproducer) or NB⁻ (HPRT⁻) mRNA.

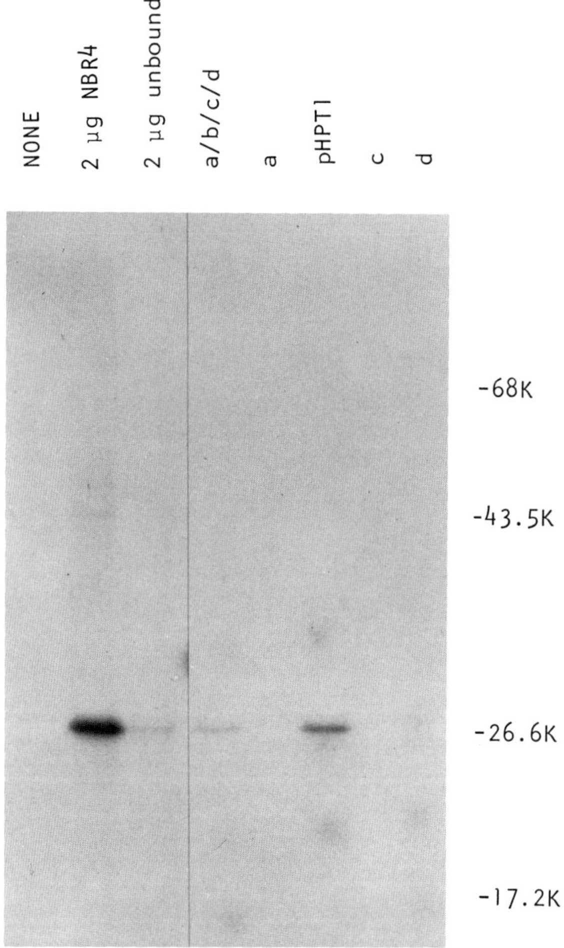

Figure 51: Radioimmune detection of (^{35}S) HPRT protein synthesized _in vitro_ from mRNA recovered after hybridization of NBR4 mRNA to plasmid DNAs.

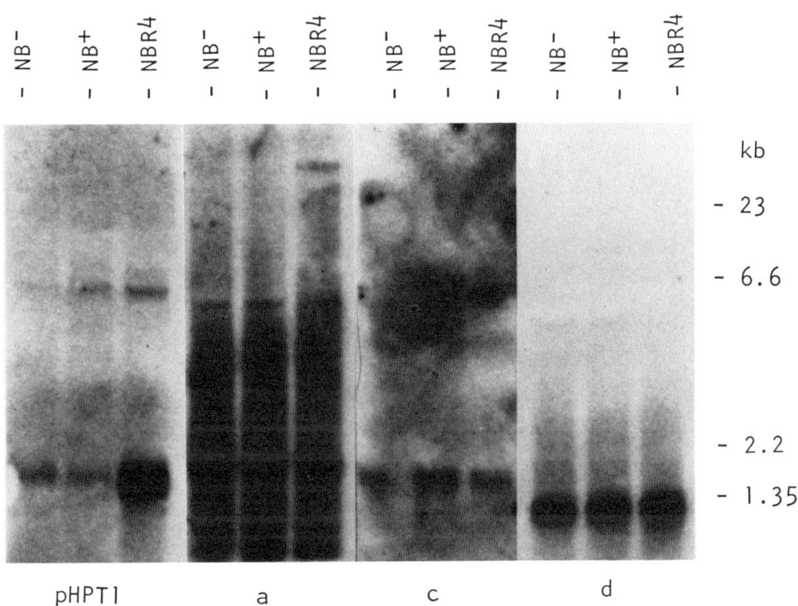

Figure 52: Hybridization of mRNA (4 µg) from NBR4 (HPRT overproducer) NB⁺ (wild type) and NB⁻ (HPRT⁻) to plasmid DNAs.

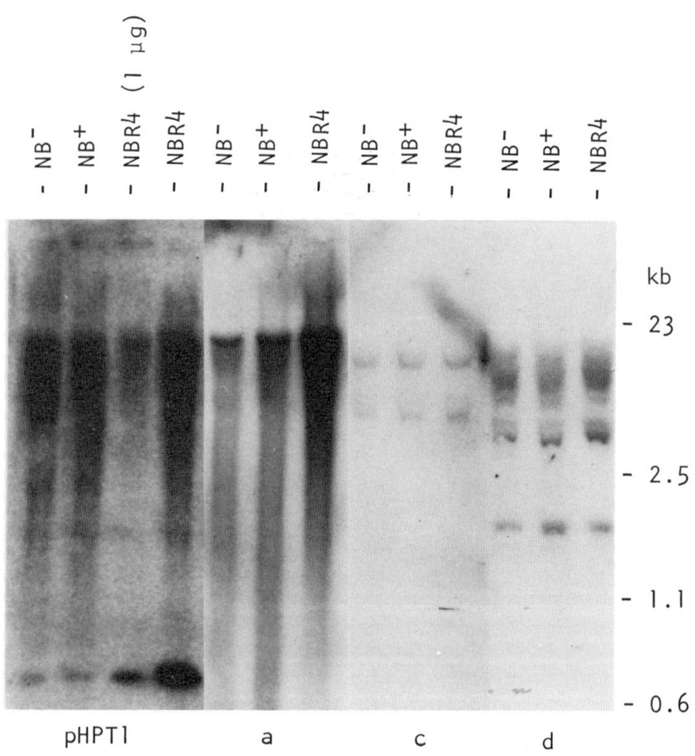

Figure 53: Southern analysis of NB⁺, NB⁻,
and NBR4 with four different cDNA recombinants.

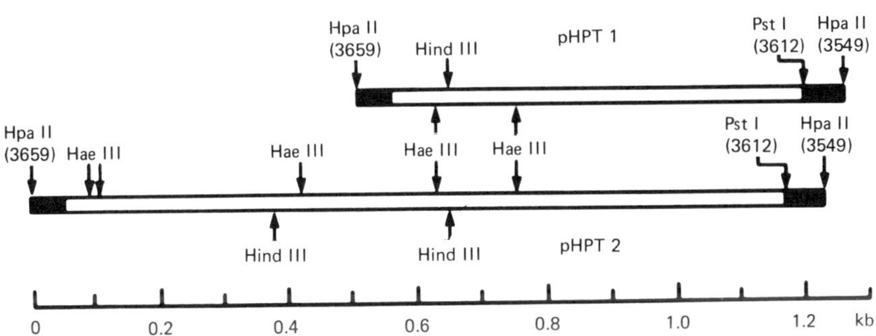

Figure 54: Restriction endonuclease cleavage sites in the recombinant plasmids pHPT1 and pHPT2.

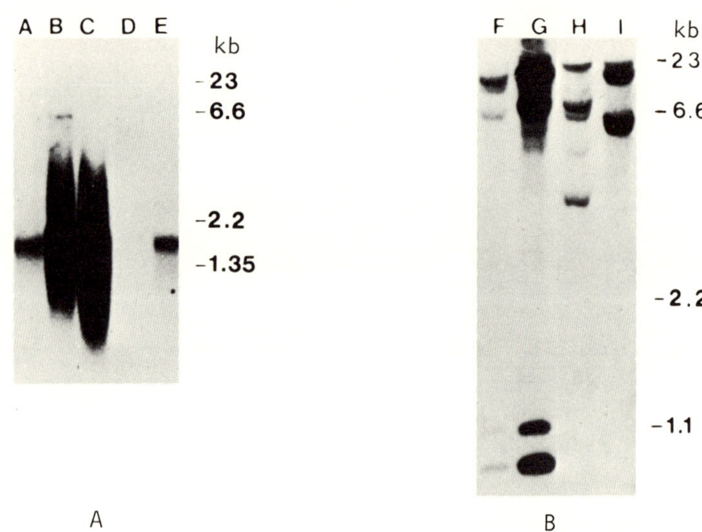

Figure 55:   Panel A corresponds to a Northern
analysis of mRNA using HPT1 as probe.   A) NB⁺
mRNA; B) NBR4 oligo dT purified mRNA; C) NBR4
sucrose gradient purified mRNA; D) Baboon mRNA;
and E) Chinese hamster.   Panel B corresponds
to a Southern analysis of DNA restricted with
Hind III. A) NB⁺; B) NBR4; C) Human; D) Hamster.

We would estimate on the basis of the size of the restriction fragments, the gene is between 20 and 30 kb using a cDNA probe of
seventy percent mRNA length. The gene appears to be significantly
larger and contains multiple introns. We know in NBR4 that not all
sequences recognized by HPT2 are amplified. Furthermore, we have
had the opportunity to study a number of spontaneous deletion mutants
of the Chinese hamster locus. They are missing bands which are
amplified in the overproducer but not those which go unamplified
in NBR4. Thus, there is a suggestion that HPRT sequences which are
not related to the active gene (pseudogene) are present in mouse
and hamster.

Before reporting a bit more of the human data, I would like to
emphasize that we have amplified the HPRT gene and other cell lines
by taking advantage of the data we had obtained on the mouse cell
line. We knew we had amplified a mutant mouse cell line. Thus,
we used hamster lines previously isolated as temperature sensitive
HPRT mutants. We carried out reversion studies under the conditional
lethal state (at 39°) for these mutants and found their preferential
mechanism for reversion was in gene amplification.

DR. FRANCKE: It seems to me that this system gives you an
excellent opportunity to compare the mechanisms of mutations in
male and female meiosis. In the collection of Lesch-Nyhan families
that I showed on Monday, there are now 11 males that are new mutants.
In addition there are 10 families in which the mother's are not
carriers. So they must have gotten that mutation from an event in
the mother's meiosis. There are ten in which the mothers are clearly
heterozygous but the maternal grandmothers are not. they could have
received their mutation from either the father or the mother. In
these cases, with increased paternal age, one could hypothesize that
it is more likely to be a paternal meiotic problem. So if you can
distinguish deletions from point mutations with this system, it
would be very interesting to look at these families.

DR. CASKEY: Dr. Francke, I think we need to do a bit more
work on developing the techniques for looking for transcriptional
alternations. But we will be able to do that shortly. Bill Nyhan
provided us a variety of Lesch-Nyhan cases. In our initial surveys
using two restriction enzymes, we have found frequent polymorphisms.

SPECIAL PROBLEMS OF POLYMORPHISMS

DR. RAY WHITE: The real objective of the recombinant DNA endeavor is to identify a segment of DNA which contains the gene of interest. Our approach uses arbitrary loci and linkage analysis because with many genetic diseases, specifically including Duchenne and the other muscular dystrophies, that is the best one can do at this point. I think the comments with respect to a need for a specific cell phenotype in order to define a DNA segment as having a specific genetic content are very much to the point. It may be possible to precisely locate a gene by deletion to saturate the region about a gene and hope to find something, although it does strike me as a rather difficult task. I must say, however, that the deletion that Dr. Francke found in Xp21 is intriguing. We are interested in that kind of approach to retinoblastoma. A consequence of having a gene as a probe doesn't mean that you give up looking for polymorphism and genetic markers. It just means that the quality of the genetic markers you get is much better. The recombination frequencies between the β-globin, for example, and the Hpa I site should be very small. No one has yet observed a recombinant despite about 1000 tests. The reason for isolating the gene is not to develop such a marker but to work out the molecular mechanism of the disease.

Let us consider linked genetic markers as an interim strategy. With enough genetic markers one can saturate the genome; that is, any arbitrary disease locus will be linked to a marker locus.

By pedigree analysis, you can develop markers for inherited diseases. These markers may be used for prenatal diagnosis. X-linkage makes the **prenatal** diagnosis easier; you don't need to track down the father. The mother and an unaffected male sib should be adequate.

Finally, for many familial diseases these methods can provide the knowledge of genotype. In the case of Duchenne muscular dystrophy for example, I think perhaps the most useful thing to do is start with a large family, identify a linked probe and then examine the question, is there heterogeneity in this disease by using genetic markers? From the extent of polymorphisms already identified one would expect a virtually infinite set of genetic markers in human DNA.

The critical questions are, how many probes, how many loci, how many polymorphic markers are required. There are some general rules that apply to mapping markers and autosomes. To observe linkage in a reasonable number of tests, a recombination fraction of 20 cM

Saturating the Human Genome with DNA Markers

(centimorgans) is needed. If we take 3200 cM for the recombination length of the human genome and evenly space our markers at a re-combination distance of 0.4 (40 cM) then it would require only 80 markers to view the human genome <u>in toto</u>. It would take four to five markers if the length of the X chromosome were 150-200 cM or it would take about 15 if the length were 600 cM, suggested by the recombination fraction between G6PD and HPRT. Of course, we do not encounter an evenly spaced set of genetic markers but instead a set of points at random. The groups at Utah and UCLA have estimated how many genetic markers must be identified and screened in order to get such an evenly spaced set. In order to make this map spaced at 0.4 (40 cM) you must start with a map spaced at 0.2 (20 cM) because the markers have to be mapped with respect to each other. For an X chromosome of genetic length, 160 cM, it would require about 20 markers, in order for 90 percent of the chromosome to be within 20 cM of the marker. For a genetic length of 600 cM, 50 markers would yield about 80 percent coverage. The number of markers needed increases disproportionately with length.

The possibility of heterogeneity in recombination over the chromosome must also be considered. Recombination is very high at telomeres which contribute a very significant fraction of the total length of 3300 cM. This consideration makes the job significantly easier, because most markers are not going to be at the telomeres. On this basis, linkages will be detected more quickly than otherwise expected.

If you have two alleles of equal frequency, 0.5, then half the women will be heterozygous. If the frequencies of the two alleles are 0.1 and 0.9, only 0.18 of the woman will be heterozygous. With more markers and more alleles the system improves. For example, with three evenly distributed alleles the frequency of heterozygotes is .67; with four alleles it is .75.

If you start with a recombinant DNA probe that sees some bit of
DNA in the genome and some number of sites which may or may not be
polymorphic, by using this same probe you can isolate adjacent
segments by virtue of their homology. By "walking," or taking steps
out to either side, the amount of DNA detected in the genome and
the number of sites that you score for polymorphism will be increased.
With respect to the β-globin locus, 60 kb of DNA has now been examined in some detail using multiple enzymes and a number of polymorphic sites have been identified. The various alleles are not
randomly disposed with respect to one another. There is substantial
linkage disequilibrium, but nonetheless, there are nine haplotypes
at a reasonable frequency. Once the linkage for a specific disease
has been discovered, the strategy changes quite dramatically. You
no longer screen but map recombinant DNA probes in order to saturate
a specific chromosomal segment of interest, whether by hybrids,
translocation deletion or _in situ_ hybridization. By this approach
a linkage of 10 or 20 cM can be reduced to a few cM. In the absence
of a specific cDNA probe, this strategy is to be preferred over
"walking." "Walking" is not practical with an initial linkage of
10-20 cM since 1 cM is equivalent to 1000 kb DNA.

     DR. LATT: Could you go through the reasoning behind the
connection between the average spacing between polymorphic probes
and the average space between any gene and any one probe?

     DR. WHITE: I can do it at one level.

     DR. LATT: For example, let us suppose your polymorphic probes
are 20 cM apart on the average. I would then assume that on the
average any given gene would be 10 cM away from the probe. If you
want to be within 10 cM on the X, and the X is 200 cM long, you
would only need 10 probes. I was getting the ratio and the direction
opposite to the one you gave.

     DR. WHITE: Let me just restate it. If we say 200 cM and
we space them at 0.4, what I said was 0.2 was the breakpoint. In
order to space them at 0.4, we need to have started with them at
0.2 so we can see how they relate.

     DR. LATT: You put probes every 40 cM?

     DR. WHITE: Yes. That is what we would like to, put probes
every 40 cM. Then any point that we drop on such a map, if evenly
spaced, would be within 20.

     DR. LATT: Then you would only need five on that hypothesis.

DR. WHITE: That is right. You don't start with them evenly
spaced and the question is how many points do you need to establish
a set that is evenly spaced. That is where the several fold factor
comes from and it depends on what level of confidence you want.

DR. LATT: Okay, but to have any given gene be on the average
within 20 cM from the probe, your probes could be 40 cM apart.

DR. WHITE: That is right. So the job is hard in the beginning
but in the end it is reasonably easy with a surprisingly small
number of probes, if our parameters are right, to describe the
human genome.

A couple of years ago we had relatively little notion of an
effective strategy in looking for DNA polymorphism. So we just
started with the biggest pieces of DNA we could find which would
be cloned segments from the Maniatis library (62). This was
work done primarily by Arlene Wyman who discovered, in her first
dozen such probes, the polymorphism shown in Figure 56, the DNA
of a panel of individuals whose DNA was digested with restriction
enzymes. Most showed two bands. We are fairly confident that each
band represents a different allele. Most everybody on this panel
was heterozygous at this locus, a consequence of eight to ten
reasonably frequent alleles. We were delighted with this first
polymorphism but have not seen any more like it. We have investigated whether the polymorphism was due to a point mutation or a
rearrangement series. It appeared to be a rearrangement series.

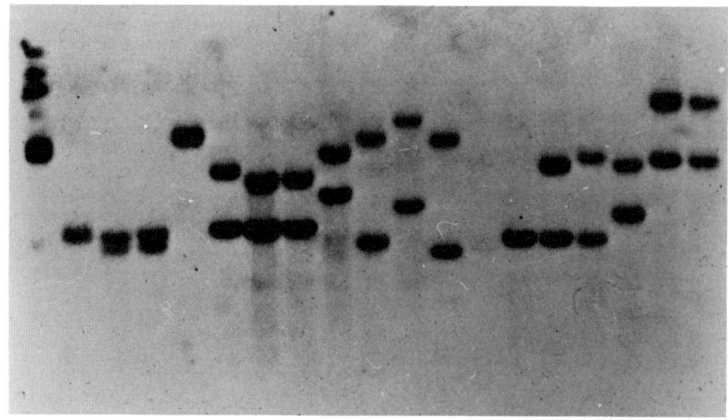

Figure 56.

When we cut with a second enzyme, the fragment lengths shifted
but the difference in length is the same, indicating a deletion
relationship between the two fragments. We have worked with an
arbitrary phage clone from the Maniatis human library in Charon 4A.
This clone was picked for this work by virtue of its not hybridizing
with human repetitious sequences. We just screened the plaques
and looked for those that are faint or negative. About half seemed
to be single copy of a length 15-18 kb. They occurred at one to
two percent frequency.

DR. DAVIDSON: Is it possible to tolerate a few Alu sequences
inserts in a 15 kb piece?

DR. WHITE: We have no way of using them as hybridization probes
and actually seeing something.

DR. HOUSMAN: I think this is a coast to coast prejudice.

DR. WHITE: I believe Dr. Housman is saying that he is also
tolerant of Alu sequences.

DR. KEDES: Is there any internally repetitive region in the
original clones?

DR. WHITE: We have not done the experiment that would answer
that question. Figure 57 uses the same probe that Dr. Francke
showed that maps on 14Q. The inheritance of such a DNA polymorphism
behaves rather analagously to protein electrophoretic variants, the
only difference being that we are looking at the length of restriction fragments as opposed to enzyme mobilities. The affected

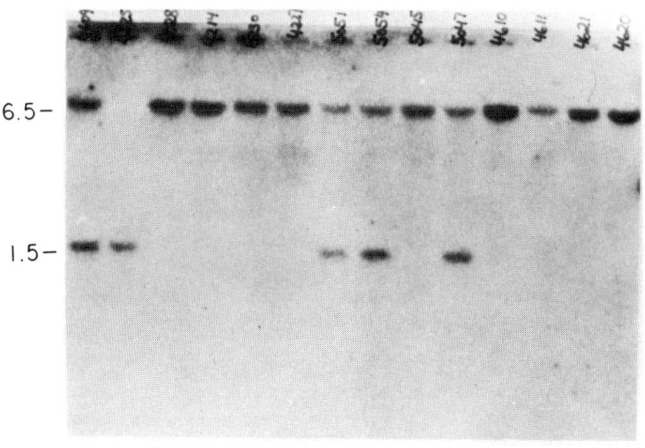

DOSLC2 / MspI

Figure 57.

individual was a member of a large breast cancer pedigree. The
mother, 4620, and the father, 4621, are both heterozygous. Four
different alleles are present. We can identify 4622 as having inherited the smaller fragment from the father and the small fragment
from the mother. The other two sibs 4623 and 4624 have inherited
reciprocal fragments.

This turns out to be quite a good locus for such things as
checking ones pedigree to make sure that you haven't switched tubes
around which occurs more frequently than I would like to acknowledge.
In this first screen done by Arlene Wyman, no base pair change, no
other polymorphisms were detected.

DR. DAVIDSON: May I ask what population you are looking at and
how long the individuals may be diverged from a single clone?

DR. WHITE: The primary reason I am in Salt Lake City is to
develop a linkage resource there with the Mormon pedigrees. Mark
Skolnik and a group of epidemiologists have constructed a computer-
ized data base that consists of the marriage, birth and death records
of over a million Mormons. They have kept excellent records over a
number of generations, have large family sizes and most are still
located in Salt Lake Valley, thus making genetic studies highly
effective.

The demography concerns about 20,000 individuals who came in
1847. They were followed by another 100,000 over the next 50 years.
Mainly, they were British, Northern European and Scandinavian in
origin. The allele frequencies in this population of various
standard isozymes and blood groups are about what you would expect
from that population, Northern European and English. they are not
a clone, they are not consanguineous and they are the group that
we are going to study, so we are defining our polymorphisms within
that set.

In combination, Mirelle Shafer and David Barker have looked
at 31 single copy clones arrived at by the same general procedure,
screening first the colonies with repetitive DNA, finding colonies
that pass that test, using those as probes for hybridization, ob-
serving that some fraction, 30-50 percent are not truly single copy
and have to be discarded, but eventually arriving at a true set of 31.
They found nine polymorphic loci in that set of tests.

Figure 57 shows DNA of individuals cut with the enzyme Msp I.
There is clear variation among the individuals. We interpret 4625
as a homozygote for the 1.5 kb fragment, 4628 a homozoygote for the
6.5 kb fragment and 4609 as a heterozygote.

In Figure 58 4611 is homozygous for the 4.3 fragment. The
two fragments, 4.3 and 12.0 have reasonably good allele frequencies,
in fact 40-60.

DR. KEDES: Can you be sure in diploid organisms that a so-
called homozygous fragment really isn't heterozygous and the other
fragment is giant and never entered the gel or never blotted?

DR. WHITE: Giant fragments do enter the gel. Our blots are
done by acid treatments. Presumably we have been transferring
uniform size DNA. Diploid genetics isn't that bad. This is an
interpretation of what we see. I will tell you a little bit more
of how we verify what we call alleles, regardless of their under-
lying structural model. This is one of David Barker's; the enzyme
is Msp I. We designate them as homozygotes for a short and longer
fragment.

Figure 59 demonstrates a polymorphism revealed by the enzyme
Taq I. In contrast to the previous polymorphism the site is
actually internal to the probe. These individuals are actually
heterozygous for a long fragment composed of two shorter fragments.
All of the polymorphisms so far appear to be base pair changes at
or near the restriction site by virtue of the fact that no
other enzyme revealed the polymorphism.

Figure 60 shows another example.

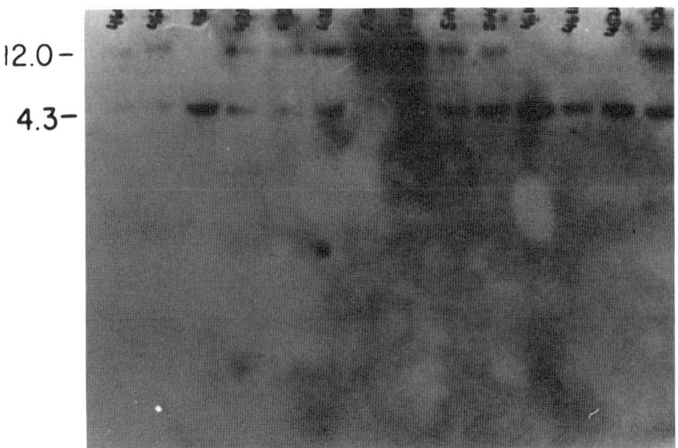

DOSLC3 / MspI

Figure 58.

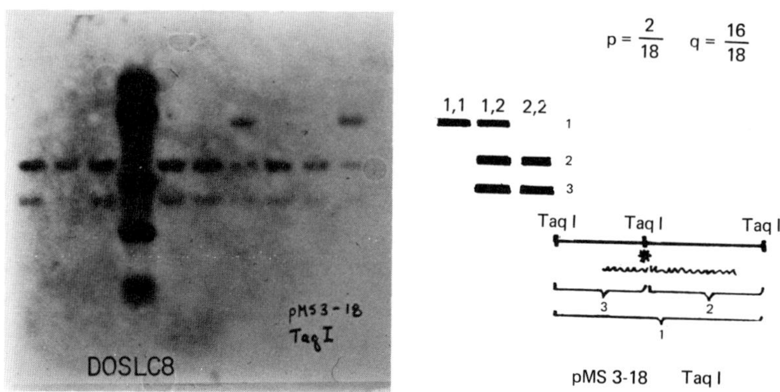

Figure   59.

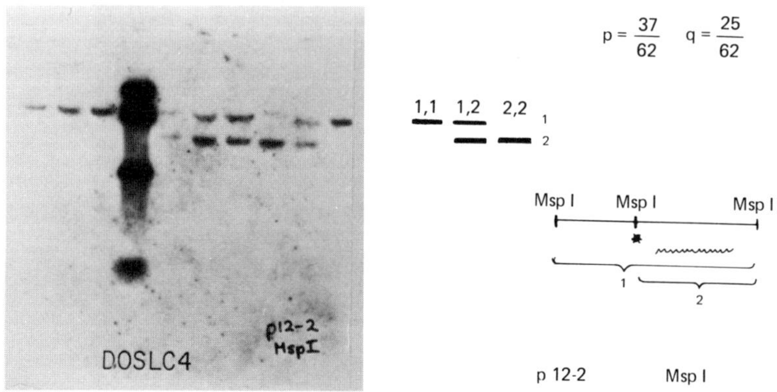

Figure   60.

Figure 61 shows a polymorphism with respect to enzyme Msp I.
The pattern is somewhat more complex because of a commonly shared
fragment in addition to the polymorphism.

Figure 62 illustrates the same locus on Figure 61 in which
the enzyme Taq I detected additional polymorphisms. Mirelle Shafer
found three Msp fragment length polymorphisms. David Barker has
found an additional six polymorphic loci, three of which were
defined with Msp I and three with Taq I. Mirelle looked with as
many as a dozen enzymes, an average of 8-10 with each probe but
found no other polymorphisms. David has found in this set a polymorphism with the enzyme EcoRI at a locus, already found to be
polymorphic by Taq I. We view that as a startling correlation of
polymorphism with the specific enzyme Msp I and Taq I that both
enzymes are useful because they show a high frequency of polymorphism.
The recognition sites of these two enzymes contain the sequence
CPG. The cytosines are very frequently methylated and as such have
been shown to be mutational hot spots. Cytosines also spontaneously
deaminate. If they are not methylated they look like uracil and are
excised and repaired. If they have methyl groups they look like
thymidines. If the mismatch is repaired or replicated a C to T
transition from the CG will result as opposed to excision of a
uracil and replacement with a cytosine. Consistent with that, if
one compares DNA sequences of the different alleles of hemoglobin
A and of insulin you find a significant portion, about 0.25 of the
single base changes are C to T transitions from CG dimer sequences.

DR. DAVIDSON: Is it obvious why U gets excised and not the G?

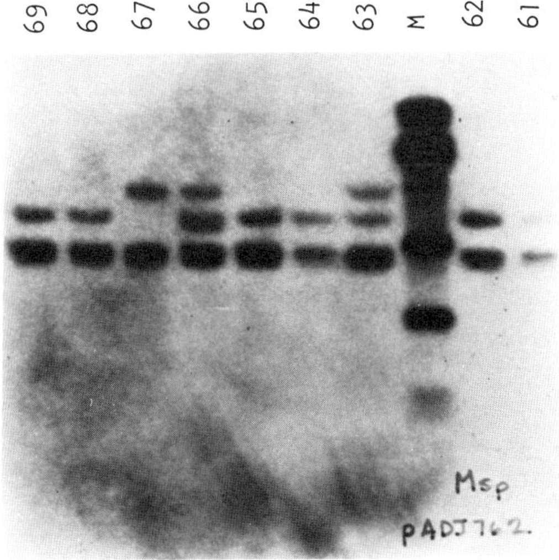

Figure 61.

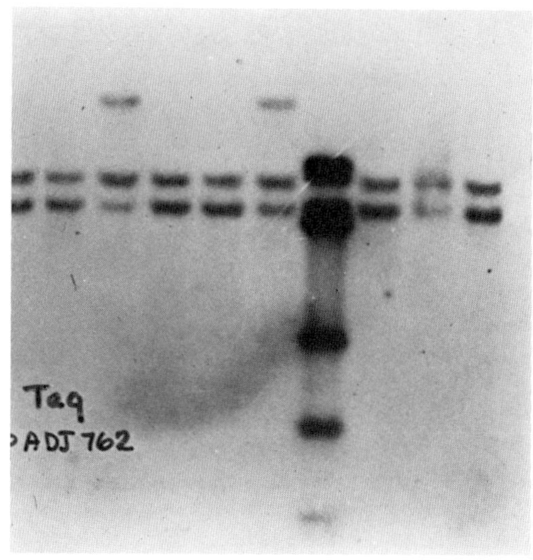

Figure 62.

        DR. WHITE: It is not obvious to me, except that there is an
enzyme which does that, which recognizes uracil in DNA and takes
it out. It is not recognizing the mismatch, but apparently it is
recognizing the uracil. I will also mention that at the time we
were sorting this out, we saw an article by Adrian Bird suggesting
that methyl cytosine should appear as hot spots in animals based
on sequence comparisons between different species (135). We feel
that the sampling we have done with restriction enzymes offers
substantial confirmation of that hypothesis.

        Several groups have isolated multiple polymorphisms within a
locus. Peter Pearson looked at a large number of different enzymes
as we did. In his comparable data
set, Taq I is elevated. There are
Screening Families for   no Msp I polymorphisms but Bgl II gave
Polymorphisms   him four or five polymorphisms. We
tested Bgl II and found 0 out of 33 loci. As far as I am concerned
this result is not understood and suggests there may be other enzymes
that are particularly useful in this kind of endeavor.

        If I were to start again looking for polymorphisms, I would
like to be able to find insertion/deletion polymorphisms that
have multiple alleles. I don't really know how to do that. I would
mention that another multiple allelic system has been described
by Graham Bell flanking the 5' end of the insulin locus (136)
and that seems to be a series based on differing numbers of a set
of tandem 15 base pairs that are repeated only in that specific

region of the genome, an interesting situation. It could, for
example be the basis for the polymorphism that we observed with
PAW 101. In seeking deletions, it is probably more useful to use
enzymes that give large fragments than to try to use large fragments
as probes. You might consider that there are actually two probes
in these kinds of experiments. There are probes for loci that you
drop on the filter, but the probes for the sites themselves are
the recipient fragments. In any event if you use the enzymes Bam HI
and Kpn I, you will look at about 20 kb of DNA, even with a very
small probe. For base pair changes, I think it is still true that
the longer the probe, the more sites you will observe. With a
polymorphism frequency of 0.1, we expect, with 5 kb probes, to see
about 0.39 of the loci polymorphic with Msp I alone. With a 10 kb
probe we would see about 0.55, which is a significant improvement.
With a 1 kb probe we would see only 0.23. It does make sense to
plan libraries that would give you fragments that are more likely
to reveal polymorphism by appropriate choice of enzymes.

Figure 63 shows what looks like a reasonably nice polymorphism.
We could say we have heterozygotes and homozygotes, possibly homozygotes for the long fragment. This was a gel done a while back
with the enzyme EcoRI. It turns out that these are in fact partial

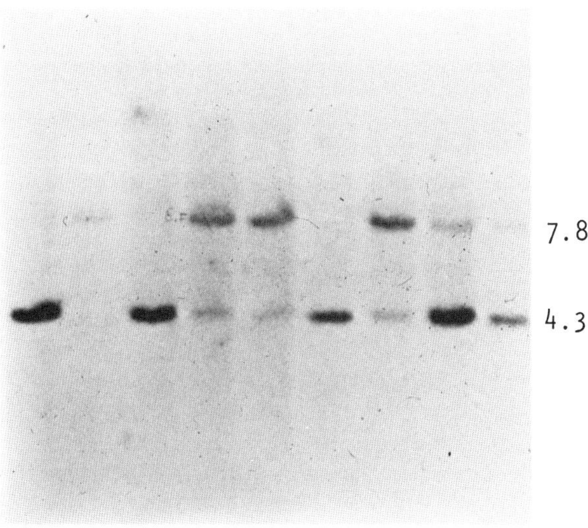

Figure 63.

digests, even though we use the standard controls for complete
digestion, mixing a little bit of λ DNA and showing that it is
totally digested. Even with the same blots, showing we have complete
digests with several EcoRI probes, this particular site turned out
to be a partial digest. We infer that some sites are more refrac-
tory than others to digest with the enzyme EcoRI. In this case,
we suspect modification of some base in the EcoRI site. There is
a cytosine at the 3' end which could be methylated. We think it is
a modification of the DNA because this same site existed in the
cloned probe, but there it seemed to have ordinary EcoRI digestion.

The only protection I think we have from artifacts in general
is showing that probes, the polymorphisms, the markers we deal with
are proper Mendelian markers. I would like to suggest that LOD
scores can be used to evaluate the correlation between a particular
polymorphic probe and a marker locus. This kind of a model comparison
doesn't prove Mendelianism, but compares the likelihood that what
you are seeing is due to Mendelianism as opposed to random.

DR. CONNEALLY: You needn't worry about using LOD scores in
this way. The man who coined the phrase, Bernard, an English
mathematician in the 1940s simply defined it as a log of the ratio
of two likelihoods, and he had no idea about linkage at the time.

DR. GUSELLA: How many of your other probes have been located
on the chromosomes localized?

DR. WHITE: Just a few.

DR. EPSTEIN: Is there any evidence for certain chromosomal
regions being relative hot spots for polymorphisms and other chromosomal regions being relatively devoid of polymorphisms? That is,
the random assumption may not be correct.

DR. WHITE: There are two levels of random. One issue is in
the picking of the probes, and as far as I know, there is no
evidence that any segment you chose for a library of single copy
probes is going to be less than arbitrary. There may be a small
fraction of fragments which are missing, but I don't think that
would bias us very much. The other issue concerns whether polymorphisms are heterogeniously distributed.

DR. BOYER: There is some information about this within genes.
The large introns of the γ-globin genes are the site of
multiple restriction length polymorphisms. In part, these polymorphisms arise because the introns are being corrected one against
the other. The introns of these two genes are about 5 kb apart
and since they are being corrected in time, one against the other,
there is a large reservoir of differences over a space of 60 bases.

Because they are continually in flux, one set of haplotypes as
opposed to the other is showing changes because the correction is
along the haplotype rather than between alleles. This naturally
makes for a reservoir of variation. It has been assumed that the
same will hold true of other such areas but I don't know of any
information about that.

DR. NADAL-GINARD: We have some information on the polymorphism
in the rat myosin heavy chain gene family. In three of eight myosin
heavy chains we have identified differences between the Sprague-
Dawley and the Wistar strain.

DR. WHITE: Are these rat strains inbred?

DR. NADAL-GINARD: I think they are scientifically outbred.
In terms of the human, we have looked now at four individuals and
none of them has the same EcoRI pattern.

DR. DAVIDSON: Do you eliminate any 15 kb insert that has an
Alu I site in it or simply any subcloned fragment?

DR. WHITE: In the beginning we eliminated any 15 kb sequences
that had an Alu I site. Subsequently, David Page has been engaged
in a kind of reconstruction, basically a subcloning, taking phages
which have only one or two Alu I sequences, cutting those with
EcoRI and reassembling them. He has found nothing startling with
respect to polymorphism; there seems to be no difference.

DR. DAVIDSON: Is that kind of discrimination selecting certain
chromosomal areas vis-a-vis others?

DR. WHITE: It is conceivable. That was part of the reason
why we subcloned EcoRI, Hind III fragments.

DR. WILLIAMSON: We have an X chromosome library from which we
have chosen about 30 clones, which appear to fall randomly on the
chromosome. We heard on the grapevine that Ray White found Msp I
and Taq I the best enzymes to use in looking for polymorphisms,
and we started with these. For our probes, Taq I has given several
polymorphisms, and Msp I, Mbo I and EcoRI have given one each.
We find approximately one single base change polymorphism per
130 base pairs screened, but some of these are only present in a
small proportion of the control population.

Locating the probes to a region of the X chromosome is a
major problem. Drs. Goodfellow, Rutter, Ropers and Francke have

been very generous with cell lines, DNA and advice, but there are
not erough cell lines with accurately mapped break points. It is
also difficult to ensure that the X-fragment is the only human DNA
in the cell line. Precise localization of cloned sequences will
be particularly important if Dr. Francke's high estimate of the
recombination length of the X chromosome is accurate, rather than
the lower estimate to which most of us have been working.

DR. NUSSBAUM: Are there any families that are informative
for one of your polymorphisms against another? Can you use them
as two linked markers?

DR. WILLIAMSON: We have only tested a single short arm clone
extensively at this time, and it does not show very close linkage
to DMD; we should have further results shortly.

DR. KEDES: If you were going to saturate the X chromosome
with these arbitrary polymorphic markers, why can't you just,
within one kindred or several kindreds, map them relative to
each other?

DR. WILLIAMSON: There are two reasons. First, it will take
a lot of time and money. Each localization, each polymorphism and
each family studied costs a lot in restriction enzymes, in isotopes
and in time. The other reason is that you need to have a family
that is polymorphic for both loci in order to map each relative to
the other. This will happen occasionally, but I feel it will be
rare, as our experience has been that even the most cooperative
family cannot always cooperate in its molecular biology. For
instance, the best DMD families that we have do not happen to be
polymorphic for the particular short arm marker we know best.

DR. HOUSMAN: I think you brought up a very important point
Dr. Williamson. I think there is a gap in the resolution between
the methods of molecular biology and the methods of classical
genetics, but I think when you look at the various strategies that
are available, there are not a lot of alternatives. I think Dr.
Francke will agree with me that _in situ_ hybridization is not very
likely to bridge that gap for you. It is just not precise enough
in terms of how close you get to where the gene is and furthermore
it is not particularly reliable for this sort of thing. Yet you need
some way of getting the fine structure map of the chromosome you
are interested in. Somatic cell genetics has a lot more promise.
There is a lot you can do and a lot has been done already. There is
a small problem in a sense that we don't have enough markers along
any chromosome, even the X, to get by with. Ideally the single copy
pieces that people are isolating might provide such markers. The

problem again is a lot of work. I think that the strategy of
Dr. Gusella talked about has the potential to give you a lot of
markers on one chromosome that will allow you to sort out a fine
structure map. I think that is going to turn out to be a very
important component, putting single copy polymorphic probes in
their place on the map and we ought to be thinking about how to
create somatic cell hybrids.

DR. JAMES GUSELLA: I would like to tell you about our
experiences with finding polymorphisms in human DNA. As Dr. White
has done we have been working with
arbitrary restriction fragments to
find polymorphisms. Our goal in the
autosomal area is to link Huntington's

Search for Polymorphisms
in Huntington's Disease

disease, so we have collected large kindreds and made permanent EBV
(Epstein Barr Virus) transformed cell lines from blood samples of
these people. In the last year or so we have made over 400 such cell
lines, of which about 200 are Huntington's and the other 200 are
from a couple of other diseases. We are now screening these DNAs
for polymorphism, obviously most of which will be autosomal. Like
Dr. White and other people here, we started trying to find polymorphisms by taking a particular single copy probe, usually chosen
only for the fact that there was no repetitive sequence in it, and
doing a Southern blot with enzyme digested DNA (several different
enzymes usually EcoRI, Hind III, Sac I) of six different individuals
who happen to come from the pedigrees of interest.

Figure 64 shows a typical result. The first six lines on
the left are Sac I digested DNA; the middle six are Hind III digested
DNA from the same individuals and the right six are EcoRI digested
DNA. The probe used was chosen out of the Maniatis library exactly
the way Dr. White has chosen his, by hybridizing to human repetitive
DNA and showing that there is none in this particular clone. There
is no difference in the Sac I pattern. The EcoRI pattern is
identical except for a very light band between the two dark bands
on top and that appears to be due to partial digestion. The Hind III
pattern, however, shows a very striking difference. In two individuals there is a band which is missing in the center of the gel and
on top there is a very light hybridization (which we think comes
from the edge of the insert in the clone) that shows a band different
in one individual from the others. We are currently characterizing
this clone in more detail. A second type of clone that we detected
is shown in Figure 65. This probe was again chosen from the Maniatis
library. This hybridization is to an EcoRI digest of DNA from
several members of one particular Huntington's family. It shows
a very small difference in an EcoRI band of about 4 kb. There
could be a 100-200 base difference between the two fragments. We
have found differences of this type fairly frequently. Out of
twelve probes we have looked at, four show small differences like
this that are very hard to resolve and are therefore not terribly

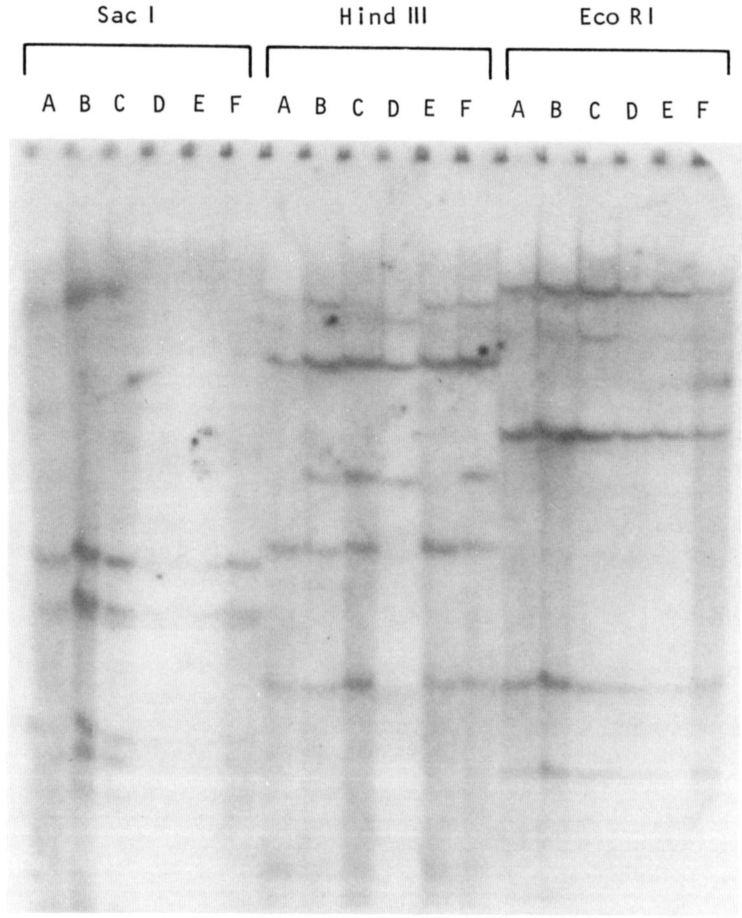

Figure 64: Hybridization of Ch4A-G8 probe to
human DNA. DNA from six indivduals (A-F) was
digested with the indicated enzyme, run on an
agarose gel, transferred to nitrocellulose
and hybridized to nick translated Ch4A G8 DNA.
Ch4A G8 is a phage selected from the Maniatis
library (137) because it contained approximately
17 kb of single copy DNA.

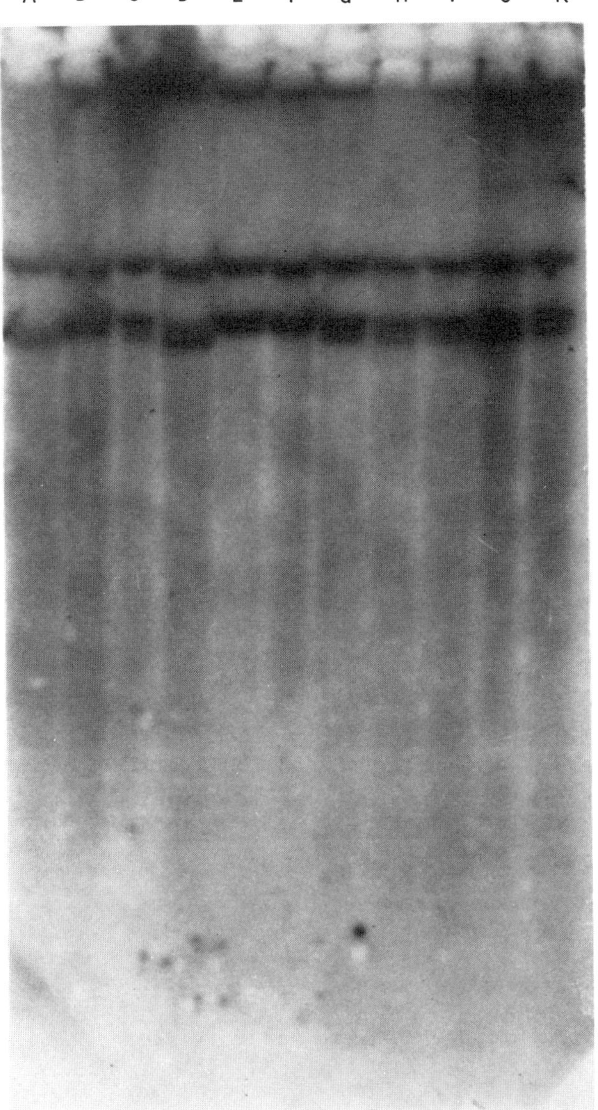

Figure 65: Hybridization of Ch4A G6 DNA to
DNA of a group of related individuals. DNA
from 11 related individuals from a Huntington's
disease pedigree was digested with EcoRI, run
on a one percent agarose gel and transferred
to nitrocellulose. The hybridization probe
used in this case, Ch4A G6 was chosen from the
Maniatis human library (62) and contained
16 kb of continuous single copy sequence.

useful unless we can find another enzyme to cut a second time that
allows us to resolve the bands a little bit better. Because of the
difficulties we have had just doing this blotting, and because of
the expense involved in digesting at least 10 micrograms of each
individual's DNA with a relatively expensive restriction enzyme, and
then having to hybridize with expensive ^{32}P-labeled probe, we decided
to devise a less expensive method which would allow us to monitor
many fragments at once.   We tried to use as probe the A36FC repetitive sequence. We hybridized it to human DNA digested with
EcoRI but unfortunately it gave a pattern that was far too complex
to resolve individual bands. We therefore worked on a method of
spreading those bands out over a larger region and, therefore, doing
somewhat of a two-dimensional analysis. What we do for example is
to resolve on an agarose gel a large amount of DNA from two individuals. We typically use a 0.8 percent gel so that the 15-20 kb DNA
is spread over a region of three cM.   We make very tight slices of
that gel so that we get exactly the same size fraction of DNA from
the two individuals. We recover the DNA from a gel, do a second
enzyme digest and run it again on a gel. We then transfer the DNA
to nitrocellulose. On this second gel we run pairs of lanes from
two individuals and each set of pairs is progressively smaller in
size (from the original enzyme digest) but has now been spread down
through the entire gel by digestion with the second enzyme. We
hybridize this filter with the repetitive sequence probe.

     Figure 66 illustrates some of the technical difficulties we
encountered in doing these gels. One typical difficulty is uneven
recovery of the DNA from the two people. In lane 1A, for example,
the DNA is clearly present whereas the recovery in lane 1B is in-
adequate for a good comparison. Another difficulty is digestion
in the second dimension. Often when the DNA comes out of an agarose
gel it is very difficult to get complete digestion of it with another
enzyme. For this particular gel the original dimension was EcoRI
and the second was Sac I. In the lanes where the DNA recovery is
about the same, most of the bands that are in one are also in the
other individual. You do appear to see a few differences but most
of the time these differences are due to partial digestion, not to
a real polymorphism. We are still working on making this technique
a bit better in terms of reproducibility of recovery and digestion,
not for the purposes of finding polymorphisms because we really don't
seem to find them in great numbers this way, but for characterizing
somatic cell hybrids where you have multiple chromosomes and these
complicated hybridization in one dimension. What this technique
has led us to, however, is another way of looking for polymorphisms
that doesn't involve Southern blotting and therefore is much cheaper
in terms of enzymes and ^{32}P.   There is no reason, once we have cut
out two fractions of DNA, that we really have to compare the fractions
by running them on a gel. We can actually make a recombinant library
of those two size fractions and that library now, instead of containing

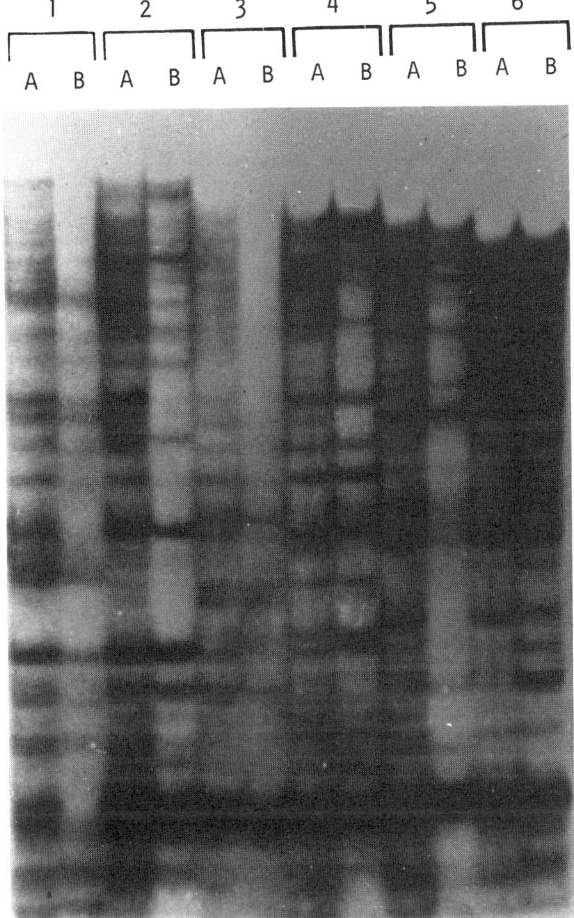

Figure 66: Comparison of DNA from two individuals
using A36FC repetitive sequence probe. DNA from two
individuals (A,B) was digested with EcoRI, and run
on an 0.8% agarose gel. The gel was then sliced to
produce identical size fractions of DNA from each
individual and the DNA was electroeluted from each
slice and recovered by ethanol precipitation. Each
fraction of DNA was then digested with Sac I, run on
a 1 percent agarose gel, and transferred to nitrocellulose. The filter was hybridized to nick translated A36FC repetitive sequence probe.

the complexity of the entire genome, may only need to contain one
or two percent of that complexity to be a complete library for that
size class. We have constructed such libraries and picked "matching"
clones out of two libraries, (that is, clones that represent essentially the homologous piece of DNA from different individuals). When
they are propagated and DNA has been prepared, the clones can be
compared directly using restriction enzymes without the need for
Southern blotting and without the need for hybridization as shown
in Figure 67. They, therefore, can be compared with many more
enzymes. Figure 67 shows the first three matching clones we
looked at. Xho I and Sal I don't cut the insert in this clone.
This middle band between the Charon 4A vector arms is the 15 kb
insert. Xba I cuts it down to several smaller fragments as does
Bgl I. Since the inserts are cloned in the same vector, we don't
have to do any kind of separations at all. We just digest with a
particular enzyme and EcoRI. The EcoRI will always liberate the
fragment and the other enzyme will create a pattern. Any difference
at all we see in this pattern is a difference in restriction site
in the insert of the recombinant phage.

The three phage clones in Figure 67 came from different people
(two of the phages come from one library and one from the other).
This technique of clone comparison has several advantages. Obviously,
it is cheaper to run the comparison for many enzymes rather than for
just two or three enzymes. It allows you to find differences that
aren't necessary in single copy DNA but that you may be able to pick
up with a single copy DNA probe. If we look at 15 kb pieces that
are completely single copy we may, therefore, bias the results.
Using this method of clone comparison we can look at any 15 kb piece
and once we find a difference, all we then have to do is find (by
mapping this particular clone) a subfragment which is single copy
DNA and which will detect the polymorphism in a Southern blot.

Obviously, in Figure 68 there are no differences. After we
looked at about a dozen enzymes with these clones we did find a
difference as seen in Figure 68, Panel A. After having used a number
of 6-base recognition sequence enzymes and finding no differences
we switched to "4 cutters." It is possible to use "4 cutters" on
the entire phage but it is a little difficult to resolve all the
bands, so the easiest way is to just cut out the 15 kb insert and
run that alone with the particular enzyme in question. In this
case the enzyme is Taq I. In HDIA and HDID there is a pattern at
the top of the gel at three bands. In HDIB the band at 3 kb is
missing. It has been replaced by two smaller bands at 1.9 kb and
1.1 kb which are not present in HDIA and HDID. HDIA and HDIB came
from the same library and therefore the prediction would be that
this individual is heterozygous for a Taq I site somewhere in
that piece of DNA.

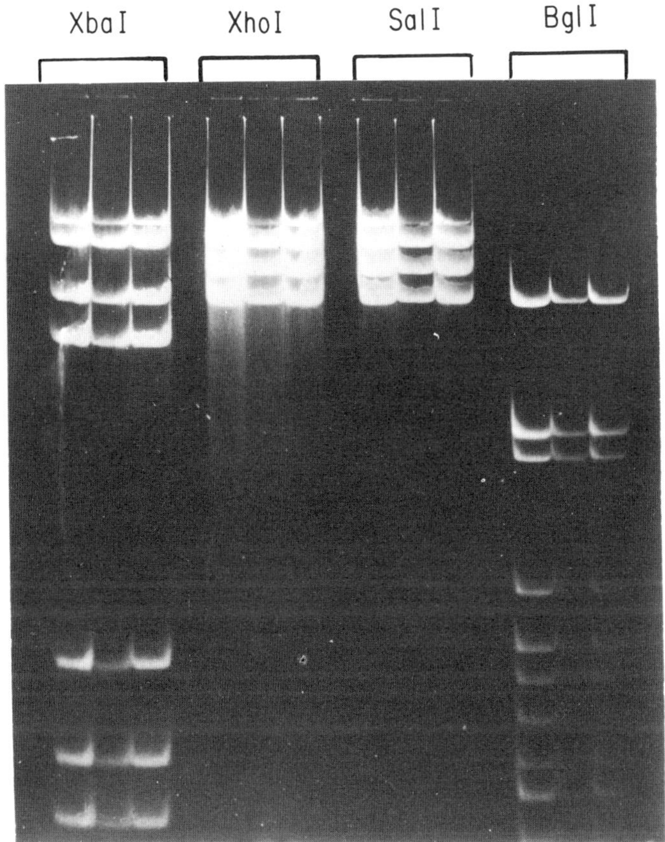

Figure 67: Clone comparison of matching clones HDIA, HDIB and HDID DNA from three matching clones (isolated as described in the text). They were digested with an indicated enzyme + EcoRI and were then compared by agarose gel electrophoresis followed by ethidium bromide staining.

Figure 68 shows a map of this clone and I should point out that
when you have a nice isolated clone like this, obtaining a map is
a matter of running one agarose gel. A map of this detail when
you are doing Southern blotting of genomic DNA is much more difficult

I have only put the Taq I sites present at the right hand end
of the phage. The others have not been mapped completely but the
variable Taq I site is marked by an asterisk. The darkened block
is a repetitive sequence which we located in the cloned DNA. Based
on the position of this Taq I site we can assume that 3.7 kb Hind III
fragment should detect the polymorphism in human DNA and without
problems of repetitive sequence.

Figure 69 shows the results of hybridizing the 3.7 kb Hind III
fragment from the same family shown in Figure 70. There are no
homozygotes for the presence of the site, but there are homozygotes
for the absence of the site. DNAs from the particular people that the
libraries were made from are in lanes A and B on this gel. The
person whom we predicted would be heterozygous (lane B) is heterozygous. The method of clone comparison allows screening for RFLPs
with many enzymes. This allows you to potentially find several
differences in the same region for which you already have a cloned
probe. These differences could form the basis of haplotypes as
Dr. White described. We have found in this HD1 sequence that within
the 3.7 kb Hind III fragment, there is an independent Hae III
difference that appears to be a single base change in the restriction
site because no other enzyme detected it. In the 1.9 kb Hind III
fragment very close to the end, there appears to be a small deletion
of perhaps 25 bases. It is very difficult to detect on a standard
agarose gel so we will probably have to go to polyacrylamide gels
to reproducibly detect it. Several enzymes detect this difference.
Clone comparison has therefore allowed us to find a clone with three
independent variations in sequence. We don't know yet whether they
are segregating independently, whether they have reached equilibrium
or whether they are still in linkage disequilibrium, but it is
potentially a very good marker. We are currently following this
approach of clone comparison for finding polymorphisms for which
we have a well characterized probe. This particular clone is just
the first clone that we have examined. We are in the process of
isolating a large number of matching clones from these libraries.
This is relatively easy to do because each library, which is only
1-2 percent of the genome, consists of perhaps 5000 phage. It is
very easy to screen for and isolate a particular clone if you have
a pool of 5000. We are about to apply this approach to Duchenne
muscular dystrophy along with Dr. Doherty of the University of
Rochester. We have collected affected and unaffected individuals
from large pedigrees for which the rest of the members will be
collected in the near future. We will isolate matching clones
specific to the X chromosome from the affected and unaffected

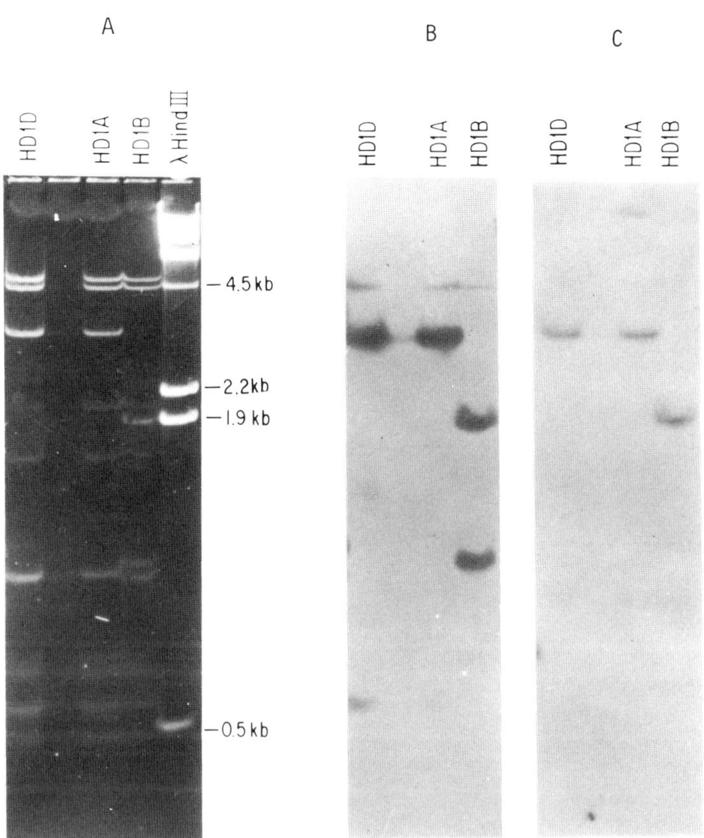

Figure 68: Clone comparison of the insert from
phages HDIA, HDIB and HDID with Taq I.
Panel A: Ethidium bromide staining agarose gel
displaying a Taq I digest of the inserts from the
indicated phages. Panel B: The gel shown in A
was Southern transferred and hybridized with a 3.7
kb Hind III fragment isolated from HDIA. This
fragment contained only single copy DNA. Panel C:
The same filter as in B was washed off and re-
hybridized to total human DNA to detect the presence
of repetitive sequences.

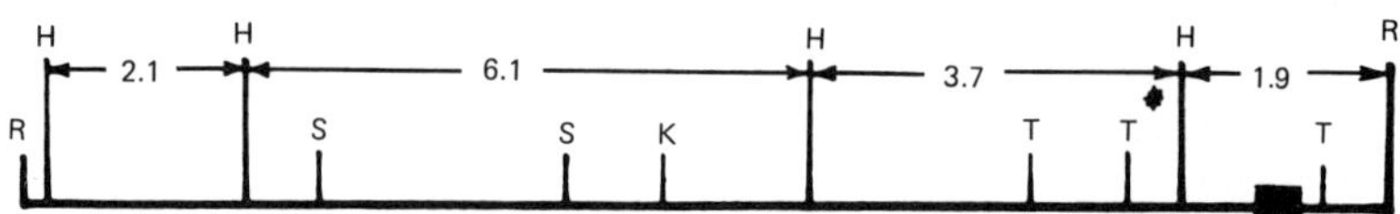

Figure 69: A restriction map of the HD1 sequence. Fragment sizes for the Hind III fragments (H) are shown in kilobases; R = EcoRI, S = Sac I, K = Kpn I, T = Taq I. Only Taq I sites within the 3.7 kb and 1.9 kb Hind III fragments are shown. The asterisk denotes the variable Taq I sites. The darkened block represents a repetitive sequence.

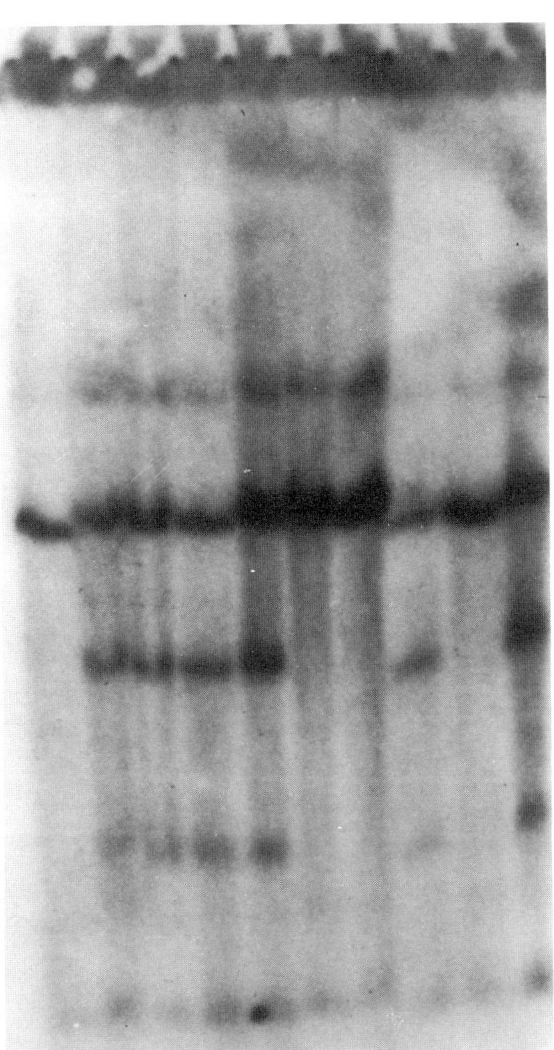

Figure 70: Demonstration of the HD1 Taq polymorphism within a pedigree. DNA from 10 related
individuals was digested with Taq I, run on a 2%
agarose gel, transferred to nitrocellulose and
hybridized to the 3.7 kb Hind III fragment from
clone HDIA shown in Figure 68.

individuals in attempts to find variants that we then will know are
segregating in the pedigree of interest. Whether they are frequent
or infrequent variants in this case doesn't particularly matter to
us because they will by definition be segregating in a large Duchenne
pedigree. I should point out that our other polymorphisms, the
autosomal ones, are not similar to Dr. Williamson's in that virtually
all of them show a very significant frequency of the variant allele.
We have not yet cloned any on the X chromosome.

With these kinds of libraries, we are not going to find the
first kind of polymorphism that Dr. White talked about, the type
that shows very large differences in size, because both of the two
matching clones won't be in the libraries. We are working on
another approach to that by making the matching libraries from a
rather wide range of pieces that are still easy to screen.

DR. NUSSBAUM: Might you not pick up one of these differences
if you are screening a large enough collection to be sure that it
should be there and you find that is a plaque that hybridizes to
your probe from one person but not from somebody else?

DR. GUSELLA: Because of the way we are making our libraries,
I think that is likely. We expect to obtain at least 5-fold as
many clones as you need to cover that complexity class. If something is not there, it either means the original size cut was not
made properly which, in general, would mean that several clones
would not be there (and that hasn't yet turned out to be the case)
or there is a real difference in size of the original EcoRI piece.
We have not found any of the latter type as yet.

DR. FRANCKE: Do you know where this one maps?

DR. GUSELLA: It maps to chromosome 5.

CHAPTER 6

STRATEGIES OF APPROACH TO
DUCHENNE MUSCULAR DYSTROPHY

DR. WILLIAMSON: Could I propose what might be called the best
and worst scenarios on a linkage to DMD, using the following
assumptions:

1.   The DMD locus is on the short arm of the X chromosome,
is a single locus (though not necessarily a single mutation), and
is probably around p21.

2.   The short arm contains somewhere in the neighborhood of
5×10^4 kb DNA, and can be "covered" completely, in linkage terms,
by ten polymorphic cloned probes.

3.   One will need about ten to twenty informative families
to establish or exclude linkage at the 10 cM level.

4.   Such a linkage, let us assume, can be localized to
approximately five percent of the short arm, or within 2,500 kb
DNA.

5.   This amount of DNA can be cloned in 50 kb cosmids in a
total of 50 clones, but about 125 will be needed to generate over-
laps; this region of the genome will also include about 100
structural (coding) genes, which should be identifiable in cDNA
libraries.

6.   This approach should allow one to identify, among DNA
sequences that are possible "candidates" for the defect, those
few structural genes that are expressed in affected tissues, and
to examine them directly.

7.   The "crossover distance" in cM during female meiosis does
not really affect the total amount of work; it means more clones
must be screened for linkage, but since the linked sequence will be
closer to the defect, fewer clones will appear as "candidates" for
the mutation causing DMD.

A panel of polymorphic X chromosome clones ordered along the
chromosome can be used, virtually unchanged, to study any X-linked
defect -- Becker dystrophy, hemophilia, Lesch-Nyhan disease and so
on. Only the family samples are changed, not the probes or the
technology.

There are a number of places where problems could arise.
Is DMD a single locus? Is it a mutation affecting an expressed
gene? If it is not, our problems will be multiplied, since we do
not understand the way in which "control" genes work, and therefore
cannot predict what changes in sequence would lead to altered
function.

DR. WHITE: Dr. Epstein asked me what one should do if having
looked at 20 or 30 markers, a linkage was not found. I said that
it would be an appropriate time to reexamine the validity of the
notion that the defect in Duchenne dystrophy is actually on the
X chromosome.

DR. SINISCALCO: Another troublesome point is the existence of
doubt that the cells which one grows from patients are meaningful
at all because they may be the result of some selection or whatever. To settle the doubt it might be worthwhile to collect cells
from all members of families in which there is a 50 percent chance
of an affected child. The cells, let us say fibroblast cultures,
could then be stored until eventually an unfortunate child is affected. At this point the cultures could be resuscitated and studied.

DR. HAUSCHKA: Some of us are trying to save, in a viable state,
the muscle cells from aborted fetuses at risk for muscular dystrophy.
Then at a time when a clearly diagnostic probe to muscular dystrophy
is found at the genetic level, one could look back at the muscle
from these fetuses to determine which ones have the disease and
then one could study those early myoblasts. Another thing I think
that would be useful in maintaining the amniocytes from all of these
fetuses. We are freezing all of those cases we can get. In the
case of newborns at risk for dystrophy, the foreskins can be saved
if they happen to be circumcized. Umbilical cords, a very good
source of fibroblasts could be saved in all such cases. These cells
will be capable of undergoing many _in vitro_ doublings and will yield
a lot of DNA. Our laboratory would be very interested in obtaining
from anyone involved in genetic counseling aborted fetuses at risk
for muscular dystrophy so that we can get the muscle tissue and

freeze it away under conditions which we know will preserve cells
in a viable state.

DR. WHITE: I am optimistic that a continued study of family
after family will yield the linkage result that we are looking for.
X chromosome linkage analysis may be readily achieved because of
the ease of determining phase.

DR. SINISCALCO: With a panel of cell hybrids with translocation
one can place with reasonable confidence something like ten markers
along the X chromosome. Human pedigrees segregating for X-linked
traits of known subregional location may also serve for subregional
mapping of RFLPs when information is available on three generations
with occurrence of recombinants. Random samples of male and female
DNAs derived from Mendelian populations should be used for the
screening of RFLPs to permit one to distinguish between true
Mendelian DNA variants and other phenomena, because only in the
first case is the frequency of a given haplotype among females
equal to the square of the frequency of the same haplotype among males.

DR. WILLIAMSON: Dr. Siniscalco has a very valuable resource
in the Sardinian cases he has characterized. However, we do not
yet know whether these persons have common polymorphisms or that
island communities are more homogeneous genetically than the people
one comes across randomly in London or New York.

DR. KUNKEL: I agree with Dr. Siniscalco. One should obtain
several clones from a particular area of the X chromosome. DNA
samples from males are best for quick identification of polymorphisms
for these cloned X probes.

DR. FRANCKE: I think at this stage of the game it would be
more efficient to put one's effort into getting as many polymorphic
loci as possible and to place them on the map with as much precision as possible before working with individual families that
have particular diseases. Getting the families DNA blotted and
put on the filter is also a lot of work. Moreover, considering
the total genetic length, if you have a limited number of probes,
chances are that you are less likely to find linkage than with
a much denser map of loci.

DR. WILLIAMSON: In Britain, we have found clinicians and
affected families very helpful; if anything, we have to emphasize
that we only want samples from selected, informative individuals
such as obligatory carrier mothers, rather than a lot of samples
which we cannot use at this time. In this way, we try to avoid
disturbing families, raising unrealistic hopes and taking many
samples from patients.

DR. SINISCALCO: Before touching the patients' families, one
has to have found the combination of probe and enzyme which yields
the best polymorphism. This can be done on random samples of males
and females, usually 100 males and 50 females. This gives us 100
X chromosomes from each sex. What you find in a Mendelian population
is likely to be true also in your family. Less homogeneous populations are more difficult to work with.

DR. SAMUEL BOYER: I disagree with Dr. Francke. I think
family studies should be pushed. In fact, families are already
being analyzed. Dr. Williamson has
shown us endonuclease-based family
studies of thyroxine-binding globulin.
That entailed a fair bit of work. I
suspect that any time someone finds

Speeding up the Search
Through Sharing and
Collaboration

an X-linked restriction-length polymorphism in the next year or so,
there is going to be a family study looking for linkage to one or
another clinical entity. I would like to make a plea for increasing
the efficiency of the hunt. We probably have enough DNA markers
already. If you enumerate them simply from descriptions given in
this room, there must be several hundred probes available or nearly
available for searching out polymorphisms along the X chromosome.
In order to use these polymorphisms effectively, what we now need
to do is to saturate the X chromosome with targets in the form of
DNA from families with X-linked disorders. At any one university
or hospital it is possible to obtain, let us say, seven or eight
concordant sib pairs with Duchenne muscular dystrophy, seven or
eight pairs with Hemophilia A and perhaps a lesser number with
Hemophilia B. After that, families with X-linked entities at most
centers become less common. As it stands, even with regard to
the more common entities, no one institution can muster a large
enough collection of suitable families to come even close to
saturating the X chromosome with targets. However, by combined
effort we might be able to do it. We could use, for the purpose,
DNA obtained either from transformed lymphoid lines or from fibroblasts. Such semi-immortal cell lines from families with X-linked
disorders could be used to form a panel with which to search for
linkage. DNA from mothers of affected individuals could serve as
a subpanel in searching for polymorphism at the outset. By
collaboration, uncommon entities, when one-by-one added to the
panel, might become sufficient for the needs of linkage analysis.
In this way, the chance that any specific restriction-length polymorphism would be found to be linked to some entity would be increased. Each investigator would thus have several targets for
his probe. Another advantage to collaboration is that we avoid
stratification and the attendant loss of information arising in
local populations where a particular polymorphism might be absent.
In order to create a panel of cell lines from which DNA can be
obtained, our motivation is a key factor. One way to motivate
contributors is to ask that everyone who wishes to use the panel

must contribute to it. At least in the beginning, if you don't
pay cell lines into the panel, you may not share in the use of
the panel. Another problem concerns reward for those who contribute
but who are not themselves involved in the business of examining
restriction-length polymorphisms. I am thinking of those neurologists,
hematologists and clinical geneticists whose patients and their relatives form the real contributors to the panel. It seems to me that
these physicians must be rewarded if their help is to be forthcoming.
One kind of reward might occur at the time of published first use
of the panel wherein everyone who contributed is named as author.
Thereafter, payment is over and the real reward for each contributor
comes when a polymorphic marker is found close to the particular
gene of their interest. While I can imagine ways of motivating
and rewarding contributors to a panel of cell lines from families
with X-linked disorders, it is still pie-in-the-sky until we find
the resources for enabling us to exchange cell lines. Simple
altruism isn't enough to enable half a dozen or so laboratories to
each obtain, grow up and transmit one or two hundred cell lines to
all other of the half dozen contributors. There must be funds for
materials and shipment, possibly through a central clearing house.
Funds will also be needed for a certain amount of record keeping
since a panel is assayed by different investigators linkage should
in time emerge between DNA markers per se. These are, of course,
the details as to what constitutes a suitable family informative
for linkage, what X-linked markers such as G6PD type are available
or can be obtained, and so on. All told, the various matters to
be worked out are considerable. However, if we don't engage in
some such collaborative enterprise as this and create a panel of
cell lines from which DNA can be obtained, the chance that any
one of us through our local resources will be able to establish
linkage between a restriction-length polymorphism and an X-linked
disease becomes quite small.

DR. EPSTEIN: That is a very important point, and there are
some parallel issues which perhaps we should discuss. For example,
it is important to discuss not only the question of patients' DNA
but also sharing the X chromosome-specific fragments that may be
used to actually analyze such patient material. There are a number
of levels at which there is already duplication.

DR. BOYER: I was asked to speak to the various groups involved
as to their willingness to share resources and participate in defining objectives. There was unanimous agreement that we should
saturate the target, the X chromosome. The questions of priorities,
details, how this is managed, all raised by a number of people.
I propose that there be created an Ad Hoc group of those who are
interested and wish to participate. The should then propose to the
MDA a mechanism whereby we can assemble pedigrees, and obtain DNA
from individuals of various X chromosomal markers. We should

concentrate on the X chromosome and not at first on any one disease entity as opposed to another so that we assemble the needed material, work out the details and then make a statement.

DR. WILLIAMSON: In my view, if the MDA or NIH were to help the groups using the different approaches to keep in touch, that would be useful. We are willing to make our probes available to those who make libraries using different techniques, to those who make cell hybrids, to those who make Goss-Harris libraries. There are a lot of resources needed, and no one group can have them all.

DR. WOLF: I just had a word with Dr. Moss. Naturally the MDA is very much interested in the potential of this sharing and he mentioned that one feasible approach would be that if someone wanted to organize such an arrangement he should get in touch with him. This might be supported on the basis of a contract with MDA.

DR. KEDES: I think what this group should think about doing is making a decision to set up a small three or four member working group to decide how best to do that and not just say leave it open ended for someone to approach the MDA. I think there should be a small committee of those who are most heavily involved to interact with MDA on doing that.

DR. HAUSCHKA: MDA requires us to fill out a disease check list at the end of our grant proposals. Analysis of Pedigrees   They could provide a similar checklist with a chromosome map on it for probes. Basically, everyone working on the system could check that off every year. MDA could just xerox them and send them to all the grantees, so everybody had everybody else's checklist.

DR. WHITE: Perhaps we should first discuss the kind of family structure that would make this kind of study most efficient. My guess is that for making a linkage map of X, having a mother, her father and 8-10 male children would probably be the most effective combination. Having her brother would not be rewarding but 6-12 male children would be. In any event I would suggest that some attention be paid to family structure. We would be delighted at Utah to cooperate in ascertaining families of various structures. We could actually do that fairly readily through the computer resources that Mark Skolnick has, if you are interested in doing that.

DR. BOYER: I wouldn't knock the grandfather and collateral method with which we have had experience for 25 years. Sib pairs, concordant sib pairs, thus eliminating fresh mutation in the mothers' meiotic divisions, afford a quick way of getting answers using conventional gels. They also serve as a source of screening for

polymorphisms in the mothers or in the sons. The sib pair method
will thus help establish whether or not one is within striking
range of a regional localization. We have recently cloned 21 sib
pairs for SS disease and were able to show five clear recombinants.
This is, of course, an autosomal condition so we don't start with
haplotypes. If a marker, perhaps an endonuclease marker, which
influences the expression of hemoglobin in the red cells were
linked but separate from the β-globin locus, we could have identified
it with essentially two gels, had we found the right enzyme at the
outset.

DR. CONNEALLY: I would like to make a pitch for larger
families. The normal sibs give you just as much information as
the affected sib. You get much more information by having larger
sibships including, of course, multiple affecteds to be sure it is
not a new mutation. Furthermore, if there is a possibility of
genetic heterogeneity we need a large sibship and, if possible,
mothers who have sisters with affecteds too.

DR. BROOKE: We have access to the families of very thoroughly
studied patients with Duchenne dystrophy. The genetics aspect of
our multiclinic study are managed by Dr. Doherty working closely
with the Boston group represented here. Not only is the diagnosis
well documented among our patients but the course of the disease
is followed and recorded in each case. The libraries obtained
from the patients will be available for use by other investigators.
We have no trouble getting material from our patients; they are
willing. We just completed the first run of bloods on the available
relatives. It took us a year to do. That was because of travel
time to collect the blood, not because the patients were unwilling.
Those patients are available to anyone in this room who wants to
utilize the material. You will have to tell us what to do. Skin
biopsies are simple; whatever you want we could get.

DR. LATT: In terms of expediting, what do you send around
cells, DNA, etc? If somebody could send around a filter with the
restricted DNA already electrophoresed and blotted, investigators
could then apply their specific probes.

DR. EPSTEIN: Possibly Dr. Doherty and Dr. Gusella have a
better way of sharing and immortalizing patients, by way of their
lymphocytes.

DR. DOHERTY: I have been working for about nine months with
Drs. Griggs and Mendell, who are part of Dr. Brooke's study and
for the past three months have been learning the technique of
immortalizing lymphocytes with EB virus in the laboratory of
Drs. Gusella and Housman. Dr. Gusella is able to do this at an
efficiency of 97 to 98 percent reliability even with samples sent
through the mail. We believe it is essential to collect the

families from the multiclinic study and have them available in a
permanent state in the form of cultures that have been frozen and
stored.

DR. WILLIAMSON: We should bear in mind that the finding of
a linkage without knowledge of the nature of the basic defect will
only have a marginal impact on the overall incidence of the disease,
since only those who already have had an affected child will (in
general) know they are at risk and come forward for antenatal
diagnosis. Obviously, this is of real value, but moving on to
find the basic defect is essential, for accurate carrier detection
and to be able to understand the mechanism that generates new
mutations leading to the disease.

DR. BROOKE: One thing geneticists should be aware of, is the
way in which clinicians are thinking about muscular dystrophy. I
am not going to talk about its

**The Elusive Pathogenesis
of Duchenne Dystrophy**

clinical course. I am going to be
very naive and totally over-simplify
the subject. The disorder is thought
to begin very early. There is a lot of speculation about whether
fetal muscle in the Duchenne patient is or is not abnormal. I
think you can find two schools of thought. There are those who
feel that normal fetal muscle is so difficult to interpret, that
interpreting muscle in the Duchenne fetus is well nigh impossible.
Others feel that they can tell the difference between Duchenne an
normal fetal muscle. Personally I belong to the camp that can't
tell the difference. In the course of this argument, the membrane
theory was proposed and that is the one that holds sway at the
moment. What we think from ultrastructural, biochemical and physiological data is that something goes wrong with the cell membrane
in the patients with Duchenne dystrophy. This may or may not lead
to an ingress of calcium into the cell which activates the lytic
process and the muscle then tries to repair that damage. It is a
constant losing battle of regeneration and degeneration, as the
muscles desperately try to heal themselves. I think I can quote
Dr. Rowland. He pointed out that everything that has been described
in the membranes in the muscle of Duchenne patients has been argued
with, transmuted or contradicted by other laboratories working in
the same area (138). I am not going to get into these arguments,
but abnormalities of membrane fluidity, phosphorylation, spectrin,
enzymes associated with the membranes have all been described and
then contradicted. Perhaps an important thing to keep in mind is
that it is very difficult to work with a Duchenne muscle. That is
the muscle is so chewed about anatomically that the findings you
come across in the Duchenne muscle are as likely to be secondary
to damage in the muscle as they are to be the primary change.

It is not just a disease of the muscle. We know there are problems with the IQ, so something must have happened to the nervous system and more interestingly, of course, the red blood cell has been implicated. That is of particular interest, because the red blood cell, as far as we can tell, is functionally normal. and you don't have to deal quite so much with the secondary structural damage. That still doesn't get you away from the problem that any change that you encounter in the blood cell may be secondary to the primary change, but at least it is not a cell that is totally chewed up as is the muscle. Please remember though that we don't know what the primary change is but most people suspect that something goes wrong with the membrane which allows leakage and that starts the process, this inevitable process of degeneration and regeneration, with degeneration eventually winning.

DR. WILLIAMSON: Is there any evidence that Duchenne muscular dystrophy has anything to do with muscle at all? I have heard it said that it is possible that the muscle cells themselves are completely normal in Duchenne and what is wrong is something that has to do with innervation.

DR. BROOKE: That was a theory that became very popular sometime ago. There were two alternate theories, the neurogenic hypothesis, holding that something in the nerve is influencing the muscle, and the vascular hypothesis suggested by the focal nature of the lesion, that there is an impairment of blood supply to the muscle. One approach that seems sensible to me is to determine which membranes are abnormal and which are normal. If we know that just the muscle, the red blood cell and neural membranes are abnormal, it would be of interest to find out what those membranes have in common rather than trying to concentrate on the whole membrane function. One way to do that, of course, is to employ a whole series of monoclonal antibodies which are specific for membrane proteins. Some of the proteins may be jointly expressed in various stages of development in the particular tissues.

DR. EPSTEIN: There is no unequivocal evidence that any red cell membrane defects in muscular dystrophy is real. That doesn't mean there are ones that are real, but the whole field is extremely equivocal. My personal opinion is that we really have very few clues about the precise defect and that is one of the reasons that those of us in the MDA Scientific Advisory Committee are so excited about the approach discussed here. It may be one of the most powerful ways of identifying the primary defect, whether myogenic or neurogenic, Dr. Brooke has presented a fair summary of what neurologists think about the disease, but whether indeed any of them are correct is still up in the air.

DR. SINISCALCO: Since there is doubt that there is a primary
defect in the red cell, hasn't anyone looked at certain heterozygotes to see whether the red cells are all abnormal.

DR. EPSTEIN: First of all, many investigators have studied
either obligate or potential carriers vs. affecteds. There is
disagreement among laboratories about these results (139).
In a review article in MUSCLE & NERVE several years ago Dr. Rowland
summarized over 20 apparent abnormalities of membranes. He cited
about an equal number of publications sayin that by a particular
experimental approach the membrane was abnormal or was normal in
Duchenne muscular dystrophy (138).

DR. CASKEY: It is possible that the actual genetic defect
in Duchenne dystrophy could be expressed in a wide variety of cell
types other than muscle? Although phenylalanine hydroxylase is
a liver enzyme, the effect of its deficiency is felt in the brain.
Adenine deaminase, an enzyme present in all cells, is related to
disease of the immune system. Hypoxanthine-guanine phosphoribosyl
transferase (HPRT) is found in virtually all cells as a housekeeping
enzyme. The target tissue there is the central nervous system.
Since we do not know the basic defect in DMD and, indeed to not
even know how to recognize a phenotype or a variant character
in DMD muscle, at the cellular level, have been led to take two
courses. The first is to try to establish in culture DMD and
and normal cells, then make direct comparisons between them. The
second is to search for a marker which could provide us an
opportunity to look for the specific gene defect responsible for
the phenotype. This objective is very approachable now.

DR. RICHARD STROHMAN: There is a great deal of ambiguity in
present biochemical studies of dystrophic muscle stemming from the
degenerative-regenerative cycle that
The Problem of Maturation   Dr. Brooke referred to. When muscle
in Muscle Cell Culture and   begins to degenerate there is a
the Phenotypic Expression   compensatory hypertrophy within the
of Duchenne Dystrophy   muscle satellite cell population.
                                 New fibers are formed from this
population as has been recently and elegantly demonstrated by both
Konigsberg and Bischoff (140-141). Our evidence from satellite cell
cultures is that the new fibers are clearly embryonic; they express
not the adult phenotype but the embryonic phenotype at least as
far as myosin heavy and light chain configuration is concerned.
In addition it is clear from the studies of Whalen in the rat and
our own studies in the chick that myosin heavy chain expression
shifts during development from an embryonic isoform to a neonatal
and then to an adult isoform. In the dystrophic muscle then one
may expect to see an adult phenotype expressed in normal fibers or
in degenerating fibers, but in the regenerating fibers we may expect
to find both embryonic and neonatal forms of myosin heavy chain.

Thus when reports appear in the literature citing myosin peptide
maps in dystrophic muscle not found in isogenic normal controls it
does not mean that unique myosin peptides are present in the
dystrophic state. The complex expression of two or three different
myosins in the dystrophic muscle resulting from the cycle of
degeneration-regeneration could easily generate myosin peptide maps
appearing to be unique. Dr. Bandman in our laboratory has been
able to generate a peptide map of myosin found in dystrophic chicken
muscle from a mixture of myosin derived from normal embryonic and
adult chicken muscle. Our evidence, therefore, is that there is
no unique dystrophic myson, but simply a complex mixture to be
expected on grounds of developmental biology.

The cycle of degeneration-regeneration is intriguing in many
ways. We know, for example, that regeneration in dystrophic muscle
is abortive. The process starts but then the new fibers also
undergo degeneration. Why? One could suppose that the new fibers
develop expressing the embryonic phenotype. Normally these fibers
would encounter an environment conducive to the transition from
embryonic to neonatal to adult expression. Suppose now that in
dystrophy that environment is unfriendly to the transition process.
Unable to synthesize the appropriate myosin, and perhaps most
certainly other isopeptides, the muscle fails to contract or fails
at some level to function appropriately and a new cycle of degeneration is initiated. Thus it appears that one avenue for a
new research direction would be one in which we look for those
conditions _in vivo_ or _in vitro_ which are essential for the transition from an embryonic to an adult phenotype expressed at the
molecular level.

Current studies in tissue culture are just beginning to
realize the extent of the limits of tissue culture approach. We,
as well as others, now know that normal tissue culture conditions
do not permit the expression of a normal adult phenotype but only
the embryonic phenotype. In the case of late onset diseases,
tissue culture studies are of marginal value without _in vitro_
conditions appropriate to adult, and therefore to late onset
disease expression.

DR. NADAL-GINARD: Preoccupation with finding a linkage and
mapping the gene must not divert us from seeking the expression of
the molecular phenotype of Duchenne dystrophy in muscle.

DR. HAUSCHKA: I agree. It would be desirable for some
laboratories to concentrate on the maturation process of normal
and dystrophic muscle cells. It would also be worthwhile for
some molecular geneticists to study DNA from myogenic cells rather
than relying exclusively on more accessible cells such as lymphocytes that, not having gone through the same developmental pathway,
may display differences, for example, in the methylation of DNA.

DR. EMERSON: Another aspect of protein regulation in muscle that we do not understand concerns the differentiation of fiber types. That problem can be viewed as a problem of selection of gene family members much as occurs in the globin situation were change from fetal to adult takes place. We simply do not know what the process is in normal muscle. Methods are now available to find out, however. The techniques of the molecular biologist should be applied to the expression of proteins in muscle during development as well as to the search for the culprit gene.

DR. EPSTEIN: I would like to reinforce the views of Drs. Strohman, Hauschka and Emerson and indicate also that the defect could be at other levels of differentiation. It is conceivable, of course, it could be some very specific myofibrillar protein that we don't know anything about yet. But I think the pattern which Dr. Brooke is more of an expert at, not only degeneration and regeneration of muscle fibers but also fibrosis and fatty infiltration, all at least suggest the possibility that you are dealing with a very basic problem of mesodermal differentiation. That is, the defect could be at the level where the mesodermal cell decides whether to be a fat cell, a muscle cell or a fibroblast. So I think we may be dealing with a different kind of disease than those attributable to a specific enzyme defect or a defect in a structural protein. Of course, it could be such, but I think we have to have an open mind that maybe we are at a new level of genetic disease.

DR. OLIVER W. JONES: The remarks made earlier this morning properly suggest that, although we have considerable interest and excitement concerning the potential of recombinant molecular biology in providing answers to unresolved problems of Duchenne muscular dystrophy, there remains the important issue of defining the molecular and clinical phenotypes for this disorder. These earlier statements also indicate there remains much needed discussion about areas of research in Duchenne muscular dystrophy other than recombinant DNA. The opportunity of constructing a summary statement of the proceedings over the past two days provides a moment of reflection on what we have learned about the X chromosome and Duchenne muscular dystrophy, with principal focus on recombinant DNA research.

I have been asked to provide an overview of the formal presentations of the past two days. Thoughts I have had during these two days were to an extent stated by others during the open discussion we have just heard. Thus, I don't think you will find my remarks original. I hope they will at least head us in the direction of formulating some of the strategies we plan to outline for future work.

From the opening discussion of prevalence and heritability by
Drs. Conneally and Brooke, we learned of the current understanding
of the clinical aspects of Duchenne muscular dystrophy. The incidence, for example, is considered now to be about 1 in 4800 male
live births. Family studies suggest that at least one-third of all
cases reflect a new mutation, and this particular point was also
supported by the data discussed briefly by Dr. Caskey and his co-
workers. Family studies suggest there is a high mutation rate at
the Duchenne locus, although Dr. Boyer raised the question as to
whether what was considered initially to be a high mutation rate
was actually so comparatively high and if I may quote him, "the
mutation rate for Duchenne dystrophy may not be as high as we
thought high actually was." Mutation frequencies also suggest a
greater rate in males over females at a ratio of about 8 to 1.
Dr. Conneally suggested this possibly could be tested by studies
of families in which there was no history of Duchenne dystrophy
but in which at least one female among several female sibs had
produced more than one male child with the disease. This is one
strategic approach which could answer this question of mutation
frequency.

Heterogeneity is thought to exist in X-linked muscular
dystrophy. Included in the data supporting heterogeneity is the
possibility of more than one gene locus involved in Duchenne
muscular dystrophy. There is evidence of a specific entity in
which mental retardation of an especially profound type is associated with certain clinical manifestations of Duchenne muscular
dystrophy. It was also suggested that Duchenne muscular dystrophy
in females is associated with structural rearrangements of the
X chromosome. Dr. Brooke presented in his discussion the concept
of the "outliers," patients with clear cut muscular dystrophy but
in whom progress of the disease is slower than ususal. These
patients could not be identified by IQ but could be identified by
tests of muscle strength and other clinical findings.

There are difficulties concerning the issue of heterogeneity
and several investigators, Drs. Brooke, Caskey, Conneally, Epstein,
Latt, Schwartz and Siniscalco, each presented differing, though
not necessarily contradictory viewpoints. Dr. Boyer offered the
alternative of "genetic modifiers" as a basis for heterogeneity
in Duchenne dystrophy. Of prime concern is the fact there is, at
present, no specific biochemical marker for Duchenne muscular
dystrophy. Moreover, we do not know very much about the only
biochemical marker available, creatine kinase (CK). Does it (CK)
really reflect Duchenne dystrophy? Age and exercise as well as
perhaps some 30 other trivial factors, affect CK, and this also
confuses the issue in Duchenne dystrophy and there is much overlap
in this particular parameter between the normal and affected
individuals and carriers. How is the CK packaged? Why is it

elevated? What does the elevation reflect? We have no answer to
any of these questions, yet these issues all interface the concept
of heterogeneity in Duchenne muscular dystrophy. Dr. Brooke
emphasized the need for and widely accepted diagnostic criteria in
the interpretation of human studies and outlined the criteria for a
clinical diagnosis of Duchenne dystrophy.

We moved on to considerations of cloning. Dr. Davidson
introduced the session by describing procedures for cloning known
genes. We learned about Alu repeats and how these are interspersed
sequences which can be used for specific probes to isolate clonal
genes. However, this procedure requires a gene for which there
is a detectable selection marker for the phenotype. At present,
there is no selection marker for Duchenne dystrophy, emphasizing
again our lack of knowledge concerning muscle cell biology in
Duchenne dystrophy.

Also, "searching" methods for gene cloning were reviewed.
Drs. Williamson and Latt independently described utilization of
cell sorting procedures for the X chromosome to initiate the
development of purification gene probes. These investigators
and their co-workers have been able to identify 5-29 clones
ranging from 5-10 kb in size. Both emphasized the importance of
X-autosome hybrid cell lines in the course of their work. The
results of these studies have yielded large clone libraries.

Dr. Kunkel reviewed efforts to regionally localize, within the
human X, DNA fragments isolated from a flow sorted X library. This
analysis utilized hybrid cells and human lines with structurally
abnormal or supernumerary X chromosomes. Dr. Bruns presented other
aspects of cloning by establishing the probes used in a mouse-human
hybrid cell model. She has studied both the X chromosome and chromosome 20, and has identified 5-6 clones on both the short and long
arm of the X chromosome.

Further discussion was devoted to review on mapping the X
chromosome. Dr. Francke began this session with an overview of
human genome mapping. Special emphasis was placed on the importance
of chromosome mutations (particularly structural rearrangements,
deletions, and duplications in the mapping process), and especially
when a chromosome marker can be found in association with a biochemically mapped marker. Based upon current knowledge, representing
approximately 10^8 base pairs, the X appears to be approximately 200
cM in length. She also emphasized that this could be an underestimate
and it might be as long as 600 cM total, with as many as 250 cM on
the short arm. She pointed out that there appeared to be some
difference in recombination frequencies between the two sexes and
also differences along particular regions of the chromosome.

A significant issue in Dr. Francke's presentation was the particular map loci for the Duchenne dystrophy gene. We were returned to the issue of expression of Duchenne dystrophy in females and the utilization of chromosome mutants for at least the tentative establishment of a locus for this particular gene. She referred to the few cases reported in which there was a translocation involving the short arm of the X chromosome at position Xp21 and reviewed the evidence from those cases that at least support the concept that the locus (or loci) is located at that site on the short arm, and also that Duchenne dystrophy may be expressed in the female. It should be emphasized, however, that the conclusion that the Duchenne dystrophy gene is located at or near Xp21 is based on the study of a small number of female patients with an X-translocation involving the Xp21 region.

Dr. Siniscalco then presented a different set of alternative approaches to mapping. One alternative involves the use of probes for subregional mapping with nucleotide regions being about 10^7 base pairs apart. The second approach utilizes pedigrees segregating for one or more of the X-linked loci. Next, Dr. Shapiro presented an innovative model for looking at X chromosome inactivation and reactivation utilizing DNA methylation probes.

On the subject of polymorphisms, Dr. White provided an excellent review of gene nucleotide sequence polymorphism and arbitrary locus mapping. The consensus seemed to be, by this point in the meeting, that, for Duchenne dystrophy, restriction fragment length polymorphisms are one of the most reasonable approaches in localizing the gene. The consensus also indicated that it is not necessary to wait for the establishment of the biochemical marker to proceed with a search for the gene, utilizing any one of the alternative approaches in recombinant DNA research. The strategy with this approach is to look for genetic markers in linkage, and everyone seemed to agree that finding a linkage was feasible prior to the identification of the structural defect.

During his discussion on DNA polymorphisms as a mechanism for mapping or developing a linkage region to the Duchenne dystrophy gene, Dr. White considered possible antenatal diagnosis of males affected with the disease. He estimated that a minimal number of three individuals in a kindred would be required to undergo analysis for polymorphism: the mother, one affected male, and an unaffected male sibling. Dr. Williamson pointed out that in his laboratory some 30 clones have been identified on the short and long arms of the X chromosome, and several demonstrate DNA polymorphisms. However, thus far no effective linkages have been identified.

After these two days of intense discussion, the consensus pointed to the search for the Duchenne muscular dystrophy gene as the most important research priority. On the other hand, with perhaps one or two exceptions, it appears that most of the laboratories involved in recombinanat DNA research do not have Duchenne dystrophy as their primary research interest. If Duchenne dystrophy proves to be a productive area for recombinant DNA technology, however, I think many laboratories will be interested in pursuing it further.

I now wish to formulate three considerations concerning the problem of Duchenne muscular dystrophy: (1) what we know with certainty, (2) what seems highly probable, and (3) what we don't know.

First of all, what do we know with certainty about Duchenne muscular dystrophy? We know it is inheritable and it is lethal. We know that certain muscular dystrophy phenotypes have all gene loci on the X chromosome, and we know that recombinant DNA technology is workable to use for definition and characterization of the gene or genes controlling muscle function.

In the category of highly probable, it seems that criteria can be established to maintain uniformity in the phenotypic characterization of Duchenne dystrophy. Recombinant DNA technology eventually may lead to an understanding of a molecular defect, carrier status, and antenatal diagnosis of hemizygotes affected with Duchenne dystrophy. I also think it is probable, but not yet proven, that one-third or more of all cases of Duchenne dystrophy reflect a new mutation and that there may be a high mutation rate at the Duchenne locus with greater frequency in males than females. Heterogeneity probably exists in this disorder. Duchenne dystrophy with mental retardation is possibly one specific expression of Duchenne dystrophy in females. The Duchenne locus or loci possibly maps on Xp21. This assignment is based upon the study of seven patients and, while likely, has not yet been proven beyond question. Anywhere from 10-15 DNA probes may be required to cover the X chromosome.

The category of unknown for Duchenne muscular dystrophy includes the biochemical phenotype, the precise genetic defect, the true mutation rate, the pathophysiologic relationship of Duchenne dystrophy to CK, and the precise length of the X chromosome.

What are the strategies that will carry our momentum in this meeting into coming months and expand the cross-fertilization of ideas which have been germinated by the sharing of research progress during the past two days? I took the liberty of extracting suggestions introduced yesterday and added a few more as a starting point for possible collaborative efforts.

One possible strategy would be to develop a panel of disease markers for the X chromosome and go directly to the family studies for DNA polymorphism rather than linkage. What would be the guidelines for participation in this effort? What would the contributor receive? What is the role of the Muscular Dystrophy Association? I think this would be a very important possibility. If this is chosen as a collaborative effort, then progress should be assessed in five years. Secondly, a pedigree bank for the linkage and mutation rate studies could be developed. The guidelines for mechanisms to be used in contracting the kindreds would have to be established.

I think it is important that the Association support establishment of criteria for the clinical diagnosis of Duchenne dystrophy, the combined use of Duchenne diagnostic criteria with CK analysis in the Duchenne dystrophy and mental retardation group, Duchenne dystrophy in the female, and CK studies within pedigrees and between pedigrees as suggested by Dr. Siniscalco. Cytogenetic studies, especially in females, should be pursued with vigor. Males with Duchenne dystrophy who are retarded and both females and males who have mental retardation and muscle disease should probably have chromosome analysis. These studies could be formulated in a nationwide collaborative effort.

CODA

The aim of the colloquium, as stated in the Preface, was
to consider which recently developed methods will most effectively
contribute to genetic studies of the human X chromosome and their
application to understanding the genetic alterations that produce
Duchenne muscular dystrophy and related neuromuscular diseases.
We already have several clues concerning the genetics of DMD. There
is clear evidence that most, if not all, cases of DMD are inherited.
This inheritance shows a typical sex-linked pattern; nearly all
cases are male; the only females who may have DMD exhibit chromosomal abnormalities. Seven females with a DMD-like disorder have
an X-autosome translocation at Xp21, suggesting that the Duchenne
locus may be in at or near this region of X short arm.

DNA from a variety of sources including human chromosomes
may be cleaved by some 100 bacterial restriction endonucleases,
each of which recognizes a specific base sequence. These cleavages
produce fragments of various lengths determined by their base
sequences and the specificity of the enzyme used. Changes in the
base sequence of the restriction sites by mutation as well as
insertions or deletions of DNA will alter the size of the DNA
fragments produced by restriction endonuclease digestion. Libraries
of cloned human X chromosome DNA have been constructed. Human X
chromosomes have been purified either by the physical method, flow
cytometry, or by biological isolation from human/rodent cell lines.
The X chromosome DNA is cleaved and inserted into an appropriate
vector, usually a drug-resistant plasmid or a modified λ phage,
that permits replication of the recombinant DNA molecule in bacteria.
In this manner, DNA segments of the human X chromosome can be cloned.

The cloned human X chromosome DNA sequences provide sensitive
probes for the detection of homologous nucleic acid sequences by
hybridization. Human DNA may be digested by a particular restriction endonuclease, and the resulting fragments can be separated by
electrophoresis. A paper replica of the electrophoretic separation
or blot can be treated with radioactively labeled DNA probe so that
only restriction fragments containing sequences homologous to probe
sequence will be detected radioautographically. Changes in the
length of restriction fragments due to mutation can be detected
by this procedure.

In the small region of the X chromosome linked to the Duchenne
locus, there is a high probability of finding one or more restriction fragment length polymorphisms, possibly within the next year
or two. Such polymorphisms will be very useful in genetic studies
of large established familial pedigrees with Duchenne cases. Tests
for the screening of female members as potential carriers and prenatal screening of fetuses within carrier mothers could be rigor-
ously established in such families.

A more long-range aim is to isolate the Duchenne gene, itself. Walking down the X chromosome or a small linkage region of the X containing about 10^6 base pairs, DNA fragment by DNA fragment, from the site of a restriction endonuclease fragment polymorphism to the altered DNA sequence responsible for DMD is within the realm of possibility. Although the consensus of discussants was that such a task would be difficult, its accomplishment would permit possible determination of the chemical alteration leading to Duchenne as well as the sequence of the normal gene. Some _in vitro_ assay for the presence of the Duchenne gene such as a phenotypic change in cultured human muscle cells would probably be necessary for the isolation of the gene.

More general probes for the detection of carriers and affected fetuses would be available. The function of the Duchenne locus might be deduced from its DNA sequence or by studies of phenotypic changes in cultured human muscle. This information is necessary for the development of effective therapy in addition to understanding the pathophysiology of DMD and related disorders.

REFERENCES

1. Jackson, C.E. and Strehler, D.A.: Limb-girdle muscular dystrophy. Clinical Manifestations and detection of preclinical disease. <u>Pediatrics</u> 41: 495-502, 1968.

2. Thompson, J.S. and Thompson, M.V.: <u>Genetics in Medicine</u> W.B. Saunders Co., Philadelphia, 1980.

3. Murphy, E.A. and Mutalik, G.S.: The application of Bayesian methods in genetic counseling. <u>Hum Hered</u> 19:126-151. 1969.

4. Haldane, J.B.S.: The rate of spontaneous mutation of a human gene. <u>J Genet</u> 31:317-326, 1935.

5. Emery, A.E.H., Skinner, R. and Holloway, S.: A study of possible heterogeneity in Duchenne muscular dystrophy. <u>Clin Genet</u> 15:444-449, 1979.

6. Prosser, E.J., Murphy, E.G. and Thompson, M.W.: Intelligence and the gene for Duchenne muscular dystrophy. <u>Clin Genet</u> 5:221-230, 1969.

7. Samaha, F.J. and Congedo, C.Z.: Two biochemical types of Duchenne dystrophy: Sarcoplasmic reticulum membrane proteins. <u>Ann Neurol</u> 1:125-130, 1977.

8. Roses, A.D., Roses, M.J., Miller, S.E., Hull, K.L. and Appel, S.H.: Carrier detection in Duchenne muscular dystrophy. <u>N Engl J Med</u> 234:193-198, 1976.

9. Pickard, N.A., Gruemer, H.D., Verril, H.L., Isaacs, E.R., Robinow, M., Nance, W.E., Myers, E.D. and Goldsmith, B.: Systemic membrane defect in the proximal muscular dystrophies. <u>N Engl J Med</u> 299:841-846, 1978.

10. Yasuda, N. and Kondo, K.: No sex difference in mutation rates of Duchenne muscular dystrophy. <u>J Med Genet</u> 17:106-111, 1980.

11. Sibert, J.R., Harper, P.S., Thompson, R.J. and Newcombe, R.G.:
     Carrier detection in Duchenne muscular dystrophy. Arch
     Dis Child 54:534-537, 1979.

12. Lubs, M.L.E., Conneally, P.M., Dumars, K.W., Greenstein, R.M.
     and Muir, W.A.: Carrier detection in Duchenne muscular
     dystrophy and implications for genetic counseling in
     X-linked disease. In: AAS Selected Symposium 33, edited
     by T.L. Sadick and S.M. Pueschel. American Association
     for the Advancement of Science, Washington, D.C. pp. 51-
     62, 1979.

13. Kuby, S.A., Keutel, H.J., Okabe, K., Jacobs, H.K., Ziter, F.,
     Gerber, D. and Tyler, F.H.: Isolation of human ATP-
     creatine transphorylases (creatine phosphokinase) from
     tissues of patients with Duchenne muscular dystrophy.
     J Biol Chem 252: 8382-8390, 1977.

14. Strickland, J.M. and Ellis, D.A.: Isoenzymes of hexokinase
     in human muscular dystrophy. Nature 253:464-466, 1975.

15. Brooke, M.H., Griggs, R.C., Mendell, J.R., Fenichel, G.M.,
     Shumate, J.B. and Pellegrino, R.J.: Clinical Trial in
     Duchenne Dystrophy, 1. The Design of the Protocol.
     Muscle & Nerve 4:186-197, 1981.

16. Lesch, M. and Nyhan, W.L.: A familial disorder of uric acid
     metabolism and central nervous system function. Am J
     Med 36:561-570, 1964.

17. Seegmiller, J.E., Rosenbloom, F.M., Kelley, W.N.: Enzyme
     defect associated with a sex-linked human neurological
     disorder and excessive purine synthesis. Science 155:
     1682-1684, 1967.

18. Migeon, B.R., Der Kaloustian, V.M., Nyhan, W.L., Young, W.J.
     and Childs, B.: X-linked hypoxanthine guanine phos-
     phoribosyltransferase deficiency: Heterozygote has two
     clonal populations. Science 160:425-427, 1968.

19. Migeon, B.R.: X-linked hypoxanthine-guanine phosphoribosyl-
     transferase deficiency: Detection of heterozygotes by
     selective medium. Biochem Genet 4:377-384, 1970.

20. Gartler, S.M., Scott, R.C., Goldstein, J.L., Campbell, B.
     and Sparkes, R.: Lesch-Nyhan syndrome: Rapid detection
     of heterozygotes by use of hair follicles. Science 172:
     572-574, 1971.

21. Francke, U., Felsenstein, J., Gartler, S.M., Migeon, B.R.,
       Dancis, J., Seegmiller, J.E., Bakay, F. and Nyhan, W.L.:
       The occurrence of new mutants in the X-linked recessive
       Lesch-Nyhan disease. Am J Hum Genet 28:123-137,
       1976.

22. Francke, U., Winter, R.M., Lin, D., Bakay, B., Seegmiller, J.E.,
       and Nyhan, W.L.: Use of carrier detection tests to
       estimate male to female ratio of mutation in Lesch-Nyhan
       disease: In: Population and Biological Aspects of Human
       Mutation (Academic Press) pp. 117-130, 1981.

23. Ricciardi, R.P., Miller, J.S., and Roberts, B.E.: Purification
       and mapping of specific mRNAs by hybridization-selection
       and cell free translation. Proc Natl Acad Sci USA
       76, 4927-4931.

24. Maniatis, T., Fritch, E.F., Lauer, J. and Lawn, R.M.: The
       molecular genetics of human hemoglobins. Ann Rev Genet
       1980.

25. Seidman, J.F., Edgell, M.H. and Leder, P.: Immunoglobulin
       light-chain structural gene sequences cloned in a bacterial
       plasmid. Nature 271:582-585, 1978.

26. Firtel, R.A.: Multigene families encoding actin and tubulin.
       Cell 24:6-7, 1981.

27. Hirsh, J. and Davidson, N.: Isolation and characterization of
       the DOPA decarboxylase gene of Drosophila melanogaster.
       Mol Cell Biol 1:475-485, 1981.

28. Snyder, M., Hirsh, J. and Davidson, N.: The cuticle genes of
       Drosophila: A developmentally regulated gene cluster.
       Cell 25:165-177, 1981.

29. Suggs, S.V., Wallace, R.B., Horose, T., Kawashima, E.H. and
       Itakura, K.: Use of synthetic oligonucleotides as hybrid-
       ization probes: Isolation of cloned cDNA sequences of
       human β_2-microglobulin. Proc Natl Acad Sci USA
       78:6613-6617, 1981.

30. Pellicer, A., Robins, D., Wold, B., Sweet, R., Jackson, J.,
       Lowy,I., Roberts, M., Sim, G.-K., Silverstein, S. and
       Axel, R.: Altering genotype and phenotype by DNA-mediated
       gene transfer. Science 209:1414-1422, 1980.

31. Lowy, I., Pellicer, A., Jackson, J.F., Sim, G.K., Silverstein, S. and Axel, R.: Isolation of transforming DNA: Cloning the hamster APRT gene. <u>Cell</u> 22:817-823, 1980.

32. Houck, C.M., Rinehart, F.P. and Schmid, C.W.: <u>J Mol Biol</u> 132:289-306, 1975.

33. Gusella, J.F., Keys, C., Varsanyi-Breiner, A., Kao, F.-T., Jones, C., Puck, T.T. and Housman, D.: Isolation and localization of DNA segments from specific human chromosomes. <u>Proc Natl Acad Sci</u> USA 77:2829-2833, 1980.

34. Murray, M.J., Shilo, B.-Z., Shih, C., Cowing, D., Hsu, H.-W. and Weinberg, R.A.: Three different human tumor cell lines contain different oncogenes. <u>Cell</u> 25:355-361, 1981.

35. Goodenow, R.S., McMillan, M., Orn, A., Nicholson, M., Davidson, N., Frelinger, J.A. and Hood, L.: Identification of a BALB/c H-2L^d gene by DNA-mediated gene transfer. <u>Science</u> 215: 677-679, 1982.

36. Botstein, D., White, R., Skolnick, M. and Davis, R.: Construction of a genetic linkage map in man using restriction fragment length polymorphisms. <u>Am J Hum Genet</u> 32:314-331, 1980.

37. Stanners, C.P., Lamm, T., Chamberlain, J.W., Stewart, S.S. and Price, G.P.: Cloning of a functional gene responsible for the expression of a cell surface antigen correlated with human chronic lymphocytic leukemia. <u>Cell</u> 27:211-221, 1981.

38. Wolf, S.F., Mareni, C.E., and Migeon, B.R.: Isolation and characterization of cloned DNA sequences that hybridize to the human X chromosome. <u>Cell</u> 21:95, 1980.

39. Schmeckpeper, B.J., Willard, J.F. and Smith, K.D.: Isolation and characterization of cloned human DNA fragments carrying reiterated sequences common to both autosomes and the X chromosome. <u>Nucleic Acids Res</u> 9:1853-1872, 1981.

40. Bruns, G.A.P., Gusella, J.F., Keys, C., Housman, D. and Gerald, P.S.: Isolation of DNA segments from the human X chromosome. <u>Pediatric Research</u> 15:559 (abstract #704) 1981.

41. Davies, K.E., Young, B.D., Elles, R.G., Hill, M.E., William-
       son, R.: Cloning of a representative genomic library
       of the human X chromosome after sorting by flow cytometry.
       Nature 293:374-376, 1981.

42. Disteche, C.M., Kunkel, L.M., Lojewski, A., Orkin, S.H.,
       Eisenhard, M., Sahar, E., Travis, B., Latt, S.A.:
       Isolation of mouse X-chromosome specific DNA from an
       X-enriched lambda phage library derived from flow sorted
       chromosomes. Cytometry, 2:282-286, 1982.

43. Kunkel, L.M., Tantravahi, U., Eisenhard, M. and Latt, S.A.:
       Regional localization on the human X of DNA segments
       cloned from flow sorted chromosomes. Accepted for
       publication, Nucleic Acids Res 1982.

44. Gusella, J.F., Keys, C., Varsany-Breiner, A., Kao, F-T.,
       Jones, C., Puck, T.T. and Housman, D.: Isolation and
       localization of DNA segments from specific human chromo-
       somes. Proc Natl Acad Sci USA 77:2829-2883, 1980.

45. Disteche, C.M., Carrano, A.V., Ashworth, L.K., Burkhart-
       Schultz, Latt, S.A.: Flow sorting of the mouse Cattanach
       X chromosome, T(X;7) 1 Ct, in an active or inactive state.
       Cytogenet Cell Genet 29:189-197, 1981.

46. Disteche, C.M., Eicher, E.M. and Latt, S.A.: Late replication
       in an X-autosome translocation in the mouse: Correlation
       with genetic inactivation and evidence for selective
       effects during embryogenesis. Proc Natl Acad Sci
       USA 76:5234-5238, 1979.

47. Latt, S.A., Willard, H.F. and Gerald, P.S.: BrdU-33258 Hoechst
       analysis of DNA replication in human lymphocytes with
       supernumerary or structurally abnormal X chromosomes.
       Chromosoma (Berl.) 57:135-153, 1976.

48. Greenstein, R.M., Reardon, M.P., Chan, T.S., Middleton, A.B.,
       Mulivor, R.A., Greene, A.E. and Coriell, L.L.: An (X;11)
       translocation in a girl with Duchenne muscular dystrophy.
       Repository Identification No. 1695. Cytogenet Cell
       Genet 27:268, 1980.

49. Latt, S.A.: Microfluorometric detection of deoxyribonucleic
       acid replication in human metaphase chromosomes. Proc
       Natl Acad Sci USA 70:3395-3399, 1973.

50. Latt, S.A.: Fluorometric detection of deoxyribonucleic
     acid synthesis; possibilities for interfacing bromo-
     deoxyuridine dye techniques with flow fluorometry.
     J Histochem Cytochem 25:913-926, 1977.

51. Benton, W.D. and Davis, R.W.: Screening lambda gt recombinant
     clones by hybridization to single plaques in situ.
     Science 196:180-182, 1977.

52. McKusick, V.A.: Mendelian Inheritance in Man. The Johns
     Hopkins University Press, Baltimore and London, 1978.

53. Carrano, A.V., Gray, J.W., Langlois, R.G., Burkart-Schultz, K.
     and Van Dilla, M.A.: Measurement and purification of
     human chromosomes by flow cytometry and sorting. Proc
     Natl Acad Sci USA, 76:1382-1385, 1979.

54. Gray, J.W., Langlois, R.G., Carrano, A.V. and Van Dilla, M.A.:
     High resolution chromosome analysis: One and two para-
     meter flow cytometry. Chromosoma 73:9-27, 1979.

55. Davies, K.E., Young, B.D., Elles, R.G., Hill, M.E. and
     Williamson, R.: Cloning of a representative genomic
     library of the human X chromosome after sorting by flow
     cytometry. Nature 293:374-376, 1981.

56. Blattner, F.R., Williams, B.G., Blechle, A.E., Denniston-
     Thompson, K., Faber, H.E., Furlong, L.A., Grunvald, D.J.,
     Kiefer, D.O., Moore, D.D., Schumm, J.W., Sheldon, E.L.
     and Smithies, D.: Charon phages: Safer derivatives
     of bacteriophage lambda for DNA cloning. Science 196:
     161-169.

57. Rigby, P.W., Dieckman, M., Rhodes, C. and Berg, P.: Labeling
     deoxyribonucleic acid to high specific activity in vitro
     by nick translation with DNA polymerase I. J Mol Biol
     113:237-251, 1977.

58. Southern, E.M.: Detection of specific sequences among DNA
     fragments separated by gel electrophoresis. J Mol Biol
     98:503-517, 1975.

59. Bolivar, F., Rodriques, R.C., Green, P.J., Betlach, M.C.,
     Heyneker, H.L., Boyer, H.W., Crosa, J.H. and Falkow, S.:
     Construction and characterization of new cloning vehicles
     II. A multi-purpose cloning system. Gene 2:95-113, 1977.

60. O'Farrell, P.: Replacement synthesis method of labeling DNA
         fragments. _Focus_ 3:1-4, 1981.

61. Bruns, G.A.P., Mintz, B.J., Leary, A.C., Regins, V.M. and
         Gerald, P.S.: Human lysosomal genes: arylsulfatase
         A and β-galactosidase. _Biochem Genet_ 17:1031-1059,
         1979.

62. Maniatis, T., Hardison, R.C., Lacy, E., Lauer, J., O'Connell,
         C., Quon, D., Sim, G.K. and Efstratiadis, A.: The
         isolation of structural genes from libraries of eukaryotic
         DNA. _Cell_ 15:687-701, 1978.

63. Yang, R.C.-A., Lis, J. and Wu, R.: Elution of DNA from agarose
         gels after electrophoresis. In: _Methods in Enzymology_
         eds. Wu, R. (Academic Press, New York) Vol. 68:176-182,
         1979.

64. Brunk, C.J., Jones, K.C. and James, T.W.: Assay for Nanogram
         Quantities of DNA in Cellular Homogenates. _Anal Biochem_
         92:487-500, 1979.

65. Jelinek, W.R., Toomey, T.P., Leinwand, L., Duncan, C.W., Biro,
         P.A., Choudary, P.V., Weissman, S.M., Rubin, C.M., Hoeck,
         C.M., Deininger, P.L. and Schmid, C.W.: Ubiguitous,
         interspered repeated sequences in mammalian genomes.
         Proc Natl Acad Sci USA 77:1398-1402, 1980.

66. Schmeckpeper, B.J., Smith, K.D., Dorman, B.P., Ruddle, F.H.
         and Conver Talbot, Jr., C.: Partial purification of
         DNA from the human X chromosome. _Proc Natl Acad Sci_
         USA 76:6525-6528, 1979.

67. Olsen, A.S., McBride, O.W. and Otey, M.C.: Isolation of unique
         sequence human X chromosomal deoxyribonucleic acid.
         Biochem 19:2419-2428, 1980.

68. Wolf, S.F., Mareni, C.E. and Migeon, B.R.: Isolation and
         characterization of cloned DNA sequences that hybridize
         to the human X chromosome. _Cell_ 21:95-102, 1980.

69. Willard, H.F. and Smith, K.D.: Identification and character-
         ization of an X-specific reiterated DNA fragment from
         the human X chromosome. Oslo Conference, Human Gene
         Mapping 6. Birth Defects: Original Article Series
         (The National Foundation, New York) Abst., 1981.

70. Yang, T.P. and Hamkalo, B.A.: Characterization of a repetitive
     DNA sequence located on the human X chromosome. J Cell
     Biol 87:107a, 1980.

71. Szabo, P. and Siniscalco, M.: Isolation and subregional
     mapping of human X chromosomal DNA probes. Oslo Conference,
     Human Gene Mapping 6. Birth Defects: Original Article
     Series (The National Foundation, New York), 1981.

72. Hill, M.E.E., Davies, K.E., Harper, P. and Williamson, R.:
     The Mendelian inheritance of a human X chromosome-specific
     DNA sequence polymorphism and its use in linkage studies
     of genetic disease. Human Genetics. In press.

73. Woods, D., Crampton, J., Clark, B. and Williamson, R.: The
     construction of a recombinant cDNA library representative
     of the poly(A)+ mRNA population from normal human
     lymphocytes. Nucleic Acids Res 5157-5168, 1980.

74. Crampton, J.M., Davies, K.E., Knapp, T.F.: The occurrence of
     families of repetitive sequences in a library of cloned
     cDNA from human lymphocytes. Nucleic Acids Res 9:3821-
     3834, 1981.

75. Solomon, E. and Bodmer, W.F.: Evolution of sickle variant
     gene. The Lancet, i:923, April 28, 1979.

76. Chang, J. and Kan, Y.W.: Antenatal diagnosis of sickle cell
     anaemia by direct analysis of the sickle mutation. The
     Lancet, ii:1127-1129, Nov. 21, 1981.

77. Geever, R.F., Wilson, L.B., Nallaseth, F.S., Milner, P.F.,
     Bittner, M. and Wilson, J.T.: Direct identification of
     sickle cell anemia by blot hybridization. Proc Natl
     Acad Sci 78:5081-5085, 1981.

78. Kazy, Z., Rozovsky, I.S. and Bakharev, V.A.: Chorion biopsy
     in early pregnancy - a method of early prenatal diagnosis
     for inherited disorders. Prenatal Diag 2:39-45, 1982.

79. New Haven Conference: First International Workshop on Human
     Gene Mapping. Birth Defects: Original Article Series,
     Vol. 10, No. 3 (The National Foundation, New York, 1974)
     Also in Cytogenet Cell Genet 13:1-216, 1973, 1974.

80. Rotterdam Conference (1974): Second International Workshop on
     Human Gene Mapping. Birth Defects: Original Article
     Series, Vol. 11, No. 3 (The National Foundation, New York,
     1975) Also in Cytogenet Cell Genet 14:162-480, 1975.

81. Baltimore Conference (1975): Third International Workshop on
     Human Gene Mapping. Birth Defects: Original Article
     Series, Vol. 12, No. 7 (The National Foundation, New
     York, 1976) Also in <u>Cytogenet Cell Genet</u> 16:1-452, 1976.

82. Winnepeg Conference (1977): Fourth International Workshop on
     Human Gene Mapping. Birth Defects: Original Article
     Series, Vol. 14, No. 4 (The National Foundation, March
     of Dimes, 1978) Also in <u>Cytogenet Cell Genet</u> 22:1-730,
     1978.

83. Edinburgh Conference (1979): Fifth International Workshop on
     Human Gene Mapping. Birth Defects: Original Article
     Series, Vol. 15, No. 11 (The National Foundation, New
     York, 1979) Also in <u>Cytogenet Cell Genet</u> 25:1-236, 1979

84. Oslo Conference (1981): Sixth International Workshop on
     Human Gene Mapping. Birth Defects. Original Article
     Series, (The National Foundation, March of Dimes, 1982)
     Also in <u>Cytogenet Cell Genet</u> 1982.

85. Cox, D.R., Francke, U. and Epstein, C.J.: Assignment of genes
     to the human X chromosome by the two-dimensional electro-
     phoretic analysis of total cell proteins from rodent-human
     somatic cell hybrids. <u>Amer J Hum Genet</u> 33:485-512, 1981.

86. Taggart, R.T. and Francke, U.: Mapping of polypeptide genes by
     two-dimensional gel electrophoresis of hybrid cell extracts.
     <u>Cytogent Cell Genet</u> 1982.

87. Vora, S. and Francke, U.: Assignment of the human gene for liver-
     type phosphofructokinase isozyme (PFKL) to chromosome 21
     using somatic cell hybrids and monoclonal anti-L antibody.
     <u>Proc Natl Acad Sci</u> USA 78:3738-3742, 1981.

88. Vora, S., Durham, S., de Martinville, B., George, D.L. and
     Francke, U.: Assignment of the human gene for muscle-type
     phosphofructokinase (PFKM) to chromosome 1 (region cen→q32)
     using somatic cell hybrids and monoclonal anti-M antibody.
     <u>Somat Cell Genet</u> 8:95-104, 1982.

89. Andres, P.W., Knowles, B.B. and Goodfellow, P.N.: A human cell
     surface antigen defined by a monoclonal antibody and
     controlled by a gene on chromosome 12. <u>Somat Cell Genet</u>
     7:435-444, 1981.

90. International System of Human Cytogenetic Nomenclature - High
     Resolution Banding. <u>Cytogenet Cell Genet</u> 31:1-32, 1981.

91. Cook, P.J.L. and Hamerton, J.L: Report of the Committee on
         the Genetic Constitution of Chromosome 1. _Cytogenet
         Cell Genet_   25:9-20, 1979.

92. Francke, U., Bakay, B., Conner, J.D., Coldwell, J.G. and
         Nyhan, W.L.: Linkage relationships of X-linked enzymes
         glucose-6-phosphate dehydrogenase and hypoxanthine guanine
         phosphoribosyltransferase: Recombination in female off-
         spring of double heterozygoes. _Am J Hum Genet_ 26:
         512-522, 1974.

93. Keats, B.J.B., Morton, N.E., Rao, D.C. and Williams, W.R.:
         A Sourcebook for Linkage in Man. Johns Hopkins University
         Press, 1979.

94. Verellen, C.H., Freund, M., DeMeyer, R., Laterre, C.H.
         Scholberg, B. and Frederic, J.: Progressive muscular
         dystrophy of the Duchenne type in a young girl associated
         with an aberration of chromosome X. _Excerpta Med Int
         Cong Ser_   425:42 (abstract), 1977. Also: _Am J Hum
         Genet_   30:97A (abstract), 1978.

95. Canki, N., Dutrillaux, B., Tivadar, J.: Dystrophie musculaire
         de Duchenne chez une petite fille porteuse d'une trans-
         location t(X;3) (p21;q13) de novo. _Ann Genet_   22:35-39,
         1979.

96. Lindenbaum, R.H., Clarke, G., Patel, C., Moncreiff, M. and
         Hughes, J.T.: Muscular dystrophy in an X;1 translocation
         female suggests that Duchenne locus is on X chromosome
         short arm. _J Med Genet_   16:389-392, 1979.

97. Greenstein, R.M., Reardon, M.P., Chan, T.S., Middleton, A.B.,
         Mulivor, R.A., Greene, A.E. and Coriell, L.L.: An (X;11)
         translocation in a girl with Duchenne muscular dystrophy.
         Cytogenet Cell Genet 27:268, 1978. Also: _Ped Res_
         11:457 (abstract) 1977.

98. Jacobs, P.A., Hunt, P.A., Mayer, M. and Bart, R.D.: Duchenne
         muscular dystrophy (DMD) in a female with an X-autosome
         translocation: Further evidence that the DMD locus is
         at Xp21. _Am J Hum Genet_ 33:513-518, 1981.

99. Emanuel, B.S., Zackai, E.H. and Tucker, S.: Further evidence
         for Xp21 location of Duchenne muscular dystrophy (DMD)
         locus: X/9 translocation in a female with DMD. _Am J
         Hum Genet_ 33:103A (abstract), 1981.

100. Vianna-Morgante, A.M., Zata, M., Campos, P. and Diament, A.J.:
     Translocation (X;6) in a female with Duchenne muscular
     dystrophy: Implications for the localization of the DMD
     locus. VI Int. Cong. Hum. Genet. Jerusalem (abstracts)
     340, 1981.

101. Vinken, P.J., Bruyn, G.W. and Ringel, S.P. (editors) Handbook
     of Clinical Neurology. Part 1, 349-414, 1979.

102. Gusella, J., Varsanyi-Breiner, A., Kao, F.T., Jones, C., Puck,
     T.T., Keys, C., Orkin, S. and Housman, D.: Precise locali-
     zation of human β-globin gene complex on chromosome 11.
     Proc Natl Acad Sci USA 76:5239-5243, 1979.

103. DiCioccio, R.A., Voss, R., Krim, J., Grzeschik, K.H. and
     Sinsicalco, M.: Identification of human RNA transcripts
     among heterogeneous nuclear RNA from man-mouse somatic
     cell hybrids. Proc Natl Acad Sci 72:1868-1872, 1975.

104. DiCioccio, R.A. and Siniscalco, M.: Human RNA transcripts in
     man-mouse somatic cell hybrids, II. Thermal denaturation
     studies and Cot analysis. Somatic Cell Genetics I:263-
     277, 1975.

105. Balazs, I., Szabo, P. and Siniscalco, M.: Hybridization prop-
     erties of human X chromosomal RNA transcripts from murine-
     human hybrids. Somatic Cell Genetics 4:617-631, 1978.

106. Balazs, I., Szabo, P. and Siniscalco, M.: Properties of human
     RNA sequences isolated from a human-mouse hybrid cell
     line. Cytogenet. & Cell Genet 22:349-351, 1978.

107. Wyman, A.R. and White, R.: A highly polymorphic locus in human
     DNA. Proc Natl Acad Sci USA 77:6754-6758, 1980.

108. Siniscalco, M., Szabo, P., Filippi, G. and Rinaldi, A.: Combin-
     ation old and new strategies for the molecular mapping of
     the human X chromosome. Proc. VIth Intl. Cong. Human
     Genet. Alan R. Liss, New York, In press.

109. Gerhard, D.S., Kawasaki, E.S., Bancroft, F.C. and Szabo, P.:
     Localization of a unique gene by direct hybridization
     in situ. Proc Natl Acad Sci USA 78:3755-3759, 1981.

110. Harper, M.E., Ullrich, A. and Saunders, G.F.: Localization of
     the human insulin gene to the distal end of the short arm
     of chromosome 11. Proc Natl Acad Sci USA 78:4458-
     4460, 1981.

111. Miller, O.J. and Siniscalco, M.: Report of the committee on
     the genetic constitution of chromosomes X and Y. Human
     Gene Mapping, 6, Oslo, 1981. Cytogenet & Cell Genet
     In press.

112. Tiepolo, L., Zuffardi, O., Fraccaro, M., DiNatale, D., Gargantini
     L., Muller, C.R. and Ropers, H.H.: Assignment by deletion
     Mapping of the steroid sulfatase X-linked ichthyosis locus
     to Xp223. Hum Genet   54:205-206, 1980.

113. Race, R. and Sanger, R.: Blood Groups in Man. Blackwell
     Scientific, Philadelphia, PA., Ch. 28, 1975.

114. Soriano, P., Szabo, P. and Bernardi, G.: The scattered
     distribution of actin genes in the mouse and human genomes
     (Submitted for publication).

115. Emery, A.E.H., Smith, C.A.B. and Sanger, R.: The linkage
     relations of the loci for benign (Becker type) X-borne
     muscular dystrophy, colorblindness and the Xg blood groups.
     Ann Hum Genet   32:261-269, 1969.

116. Rudak, E., Mayer, M., Jacobs, P., Sprenkel, J., Do, T. and
     Migeon, B.: X-11 translocation:replication and mapping
     studies. Human Gene Mapping 5: Cytogenet & Cell Genet
     25:199-200, 1979.

117. DeMeyer, R., Verellen, C., Freund, M., Laterre, C., Scholberg,
     B. and Frederic, J.: Progressive muscular dystrophy
     of the Duchenne type in a young girl associated with an
     aberration of chromosome X. In: Excerpta Medica,
     Amsterdam-Oxford, Littlefield, Birth Defects, Montreal,
     Canada, p. 42, 1977.

118. Verellen, C., Markovic, B., DeMeyer, R., Freund, M., Laterre, C.
     and Worton, R.: Expression of an X-linked recessive
     disease in a female due to non-random inactivation of
     the X chromosome. Am J Hum Genet   30:97A, 1978.

119. Persico, M.G., Toniolo, D., Nobile, C., D'Urso, M. and
     Luzzato, L.: cDNA sequences of human glucose-6-
     phosphate dehydrogenase cloned in pBR322. Nature 294:
     778-780, 1981.

120. Szabo, P., Soriono, P. and Davatelis, G.N.: Localization of
     an actin gene on the human X chromosome. FASEB 1982
     In press.

121. Mohandes, T., Sparkes, R.S., Hullkuhl, B., Grzeschik, K.H. and Shapiro, L.J.: Expression of an X-linked gene from an inactive human X chromosome in mouse-human hybrid cells: Further evidence for the non-inactivation of the steroid sulfatase locus in man. *Proc Natl Acad Sci* USA, 77:6759-6763, 1980.

122. Mohandas, T.K., Shapiro, L.J., Sparkes, R. and Sparkes, M.C.: Regional assignment of the steroid sulfatase-X-linked ichthyosis locus. Implications for a non-inactivated region on the short arm of the human X chromosome. *Proc Natl Acad Sci* USA, 76:5779-5783, 1979.

123. Shapiro, L.J., Mohandes, T., Weiss, R. and Romeo, G.: Non-inactivation of an X chromosome locus in man. *Science* 204:1224-1226, 1979.

124. Kahan, B. and DeMars, R.: Localized derepression on the human inactive X chromosome in mouse-human cell hybrids. *Proc Natl Acad Sci* USA, 72:1510-1514, 1975.

125. Hellkuhl, B. and Grzeschik, K.H.: Partial reactivation of a human inactive X chromosome in human-mouse somatic cell hybrids. *Cytogenet Cell Genet* 22:527-530, 1978.

126. Riggs, A.: X-inactivation, differentiation, and DNA methylation. *Cytogenet Cell Genet* 14:9-25, 1975.

127. Holliday, R. and Pugh, J.E.: DNA modification mechanisms and gene activity during development. *Science* 187:226-232, 1975.

128. Constantinides, P.G., Jones, P.A. and Gevers, W.: Functional straited muscle cells from mon-myoblast precursors following 5-azacytidine treatment. *Nature* 267:364-366, 1977.

129. Jones, P.A. and Taylor, S.M.: Cellular differentiation, cytidine analogues and DNA methylation. *Cell* 20:85-93, 1980.

130. Mohandes, T., Sparkes, R.S. and Shapiro, L.J.: Reactivation of an inactive human X chromosome: Evidence for X-inactivation by DNA methylation. *Science* 211:393-396, 1981.

131. Bishop, D.F., Kovac, C.R. and Desnick, R.J.: Enzyme therapy
         XX: Further evidence for the differential in vivo fate
         of human splenic and plasma forms of α-galactosidase A
         in Fabry disease, in lysosomes and lysosomal storage
         diseases, edited by J.W. Callahan and J.A. Lowden,
         381-394, Raven Press, New York, 1981.

132. Venolia, L., Gartler, S.M. Wassman, E.R., Mohandas, T. and
         Shapiro, L.J.: Transformation with DNA from 5-azacytidine
         reactivated X chromosomes. Proc Natl Acad Sci
         79:2352-2354, 1982.

133. Jones, P.A., Taylor, S.M., Mohandes, T. and Shapiro, L.J.:
         Cell cycle specific reactivation of an inactive X chromo-
         some locus by 5-azacytidine. Proc Natl Acad Sci
         79:1215-1219, 1982.

134. Razin, A. and Riggs, A.D.: DNA methylation and gene function.
         Science 210:604-610, 1980.

135. Bird, A.P.: DNA methylation and the frequency of CpG in
         animal DNA. Nucleic Acids Research 8:1499-1504,
         1980.

136. Bell, G.K., Celby, M.J., and Rutter, W.J.: The higly poly-
         morphic region near the human insulin gene is composed
         of simple tandemly repeating sequences. Nature Vol. 295,
         31-35, 1982.

137. Laun, R., Fritsch, E., Parker, R.C., Blake, G. and Maniatis, T.:
         The isolation and characterization of linked delta and
         beta-globin genes from a cloned library of human DNA.
         Cell 15:1157-1174, 1978.

138. Roses, A.D., Emery, A.E.H., Harper, P.S., Thompson, M.W.
         and Walton, J.: Carrier Detection Panel Discussion.
         In: Disorders of the Motor Units. Schotland, D.L.
         (editor) 847-860, Wiley, New York, 1981.

139. Rowland, L.T.: Biochemistry of muscle membranes in Duchenne
         muscular dystrophy. Muscle & Nerve 3:3-20, 1980.

140. Konigsberg, W.R., Lipton, B.H., Konigsberg, I.R.: Developmental
         Biology 45:260-275, 1975.

141. Bischoff, R. et al: Plasticity of Muscle, Pette, D. (editor)
         De Gruyter Press, 119-129, Berlin, 1979.

INDEX